T0140392

Springer Theses

Recognizing Outstanding Ph.D. Research

Aims and Scope

The series "Springer Theses" brings together a selection of the very best Ph.D. theses from around the world and across the physical sciences. Nominated and endorsed by two recognized specialists, each published volume has been selected for its scientific excellence and the high impact of its contents for the pertinent field of research. For greater accessibility to non-specialists, the published versions include an extended introduction, as well as a foreword by the student's supervisor explaining the special relevance of the work for the field. As a whole, the series will provide a valuable resource both for newcomers to the research fields described, and for other scientists seeking detailed background information on special questions. Finally, it provides an accredited documentation of the valuable contributions made by today's younger generation of scientists.

Theses are accepted into the series by invited nomination only and must fulfill all of the following criteria

- They must be written in good English.
- The topic should fall within the confines of Chemistry, Physics, Earth Sciences, Engineering and related interdisciplinary fields such as Materials, Nanoscience, Chemical Engineering, Complex Systems and Biophysics.
- The work reported in the thesis must represent a significant scientific advance.
- If the thesis includes previously published material, permission to reproduce this must be gained from the respective copyright holder.
- They must have been examined and passed during the 12 months prior to nomination.
- Each thesis should include a foreword by the supervisor outlining the significance of its content.
- The theses should have a clearly defined structure including an introduction accessible to scientists not expert in that particular field.

More information about this series at http://www.springer.com/series/8790

Neil David Barrie

Cosmological Implications of Quantum Anomalies

Doctoral Thesis accepted by
the University of Sydney, Sydney, NSW, Australia

 Springer

Author
Dr. Neil David Barrie
Kavli Institute of Physics
  and Mathematics of the Universe
The University of Tokyo
  Kashiwa Campus
Kashiwa, Japan

Supervisor
Asst. Prof. Archil Kobakhidze
The University of Sydney
Sydney, NSW, Australia

ISSN 2190-5053   ISSN 2190-5061   (electronic)
Springer Theses
ISBN 978-3-030-06904-9   ISBN 978-3-319-94715-0   (eBook)
https://doi.org/10.1007/978-3-319-94715-0

Printed on acid-free paper

This Springer imprint is published by the registered company Springer International Publishing AG
part of Springer Nature
The registered company address is: Gewerbestrasse 11, 6330 Cham, Switzerland

Supervisor's Foreword

The understanding of today's large and complex universe is impossible without understanding the 'initial conditions' that led to the large-scale structures. The initial data were set in the very early universe when the quantum effects were profound. This fascinating relation between microscopic quantum laws of nature and the observable features of the universe is the topic of Neil Barrie's Ph.D. Thesis.

The introductory chapter of the thesis provides a concise description to the standard model of particle physics—up to now, the most accurate microscopic theory of physics at smallest scales—and to the standard model of hot Big Bang cosmology—the most credible theory of physics at largest scales. It also contains a critical overview of unexplained phenomena, such as horizon and flatness problems, the origin of the matter–antimatter asymmetry and mass hierarchy problem, together with a description of paradigm of cosmic inflation. The original research is presented in the remaining chapters.

In Chap. 2, the author proposes a new model of inflation based on nonlinearly realised scale invariance. The salient feature of the model is the existence of a flat direction in a generic scalar potential, which is lifted quantum mechanically, due to the quantum scale anomaly and lead to a successful inflationary era in the early universe. In Chaps. 3 and 4, two distinct scenarios for the dynamical generation of matter–antimatter asymmetry during inflation are proposed. The first is based on the dynamics of a new gauge boson with quantum anomaly, while the second utilises the ratchet mechanism. In Chap. 5, it is argued that homogeneous cosmic neutrino background, which is one of the unambiguous predictions of the standard Big Bang cosmology, develops instability in the presence of nonzero Lepton number, due to the mixed gravity-lepton number quantum anomaly. Chapter 6 is reserved for conclusions, while useful formulae and further technical details are collected in appendices.

The topics studied in Neil Barrie's thesis cover some of the most profound questions about our universe. I believe this work will be useful for students and experienced researchers interested in this area of fundamental science.

Sydney, Australia
May 2018

Asst. Prof. Archil Kobakhidze

Abstract

The aim of this thesis is to investigate the possible implications of quantum anomalies in the early universe. We first consider a new class of natural inflation models based on scale invariance, imposed by the dilaton. In the classical limit, the general scalar potential necessarily contains a flat direction; this is lifted by quantum corrections. The effective potential is found to be linear in the inflaton field, yielding inflationary predictions consistent with observation.

A new mechanism for cogenesis during inflation is presented, in which a new anomalous $U(1)_X$ gauge group is introduced. Anomaly terms source $\mathcal{CP}$ and X violating processes during inflation, producing a nonzero Chern-Simons number density that is distributed into baryonic and dark matter. The two $U(1)_X$ extensions considered in this general framework, gauged B and $B-L$ each containing an additional dark matter candidate, successfully reproduce the observed parameters.

We propose a reheating Baryogenesis scenario that utilises the ratchet mechanism. The model contains two scalars that interact via a derivative coupling, an inflaton consistent with the Starobinsky model, and a complex scalar baryon with a symmetric potential. The inflaton-scalar baryon system is found to act analogously to a forced pendulum, with driven motion near the end of reheating generating an η_B consistent with observation.

Finally, we argue that a lepton asymmetric cosmic neutrino background develops gravitational instabilities related to the mixed gravity-lepton number anomaly. In the presence of this background, an effective Chern Simons term is induced which we investigate through two possible effects, namely birefringent propagation of gravitational waves and the inducement of negative energy graviton modes in the high-frequency regime. These lead to constraints on the allowed size of the lepton asymmetry.

These models demonstrate that a concerted approach in cosmology and particle physics is the way forward in exploring the mysteries of our universe.

Acknowledgements

First and foremost, I would like to thank my supervisor Archil Kobakhidze for introducing me to the world of particle physics and for giving me the tools to explore the field. I have thoroughly enjoyed my Ph.D. studies and have grown considerably as a researcher under your guidance. I look forward to continuing our work together into the future. I would also like to thank my associate supervisor Michael Schmidt for our many fruitful discussions over the course of my studies.

Thank you to all my collaborators for the opportunity to work with you and for all the many insightful exchanges we have shared. I hope we shall continue to work together in the years to come. Special thanks to Kazuharu Bamba, Akio Sugamoto, Tatsu Takeuchi, and Kimiko Yamashita, for letting me join in their research pursuits. Also, a big thank you to Matthew Talia and Lei Wu for the many useful discussions over the course of our collaborations.

To the residents of Room 342, thank you for the many informative and humorous discussions. These last few years would not have been as enjoyable or as filled with laughter without you. Particular thanks go to my long-time office mates—Adrian Manning, Carl Suster, Matthew Talia, Jason Yue.

I would also like to acknowledge Shennong for his amazing discovery 5000 years ago, which has made the last few years highly enjoyable and almost stress-free.

To Sonia, thank you for always being so supportive, positive and understanding. You have helped me more than you know during the stressful final half of my degree.

Finally, thank you to my family and friends for their ongoing support and help over the last few years, without which none of this could have been possible.

Contents

Chapter 1
Introduction

The Standard Models of Particle Physics and Cosmology have been highly successful at describing and reproducing the observed dynamics and properties of the Universe, but they are incomplete. Many mysteries regarding the workings of nature are yet to be resolved, for which new physics beyond the standard paradigms is required. Examples of these are the properties of neutrinos, the identity and origin of dark matter and dark energy, the origin of the matter-antimatter asymmetry, the inflationary mechanism, the quantum nature of gravity, the hierarchy problem, and more; each of which are indications that physics beyond the Standard Model exists. In the past few decades, many extensions to the Standard Model have been postulated in an attempt to explain and provide solutions to these problems. Many of these models have tried to solve the various problems of the Standard Model simultaneously. Any such extensions normally have many phenomenological implications, which can allow for the utilisation of a variety of tools to constrain the models. As of yet, none of these extensions have been accepted because they either are ruled out by experimental searches or current experiments are not sensitive enough to exclude their predictions.

It is no coincidence that the mysteries of the Standard Model of Particle Physics are predominantly associated with the very early universe and its evolution, rather this is the result of a strong intertwining of cosmological and particle dynamics. Despite the apparent divide between the cosmological and particle scales today, they are intimately connected, with potential discoveries in either field having large ramifications on the other. Therefore, to gain a deeper understanding of the primordial evolution of the universe it is imperative to understand the properties of the fundamental particles of nature, given that at very early times the microscopic dynamics of these particles directly dictated this evolution. The interconnectedness of these two fields is particularly relevant to modern day theoretical explorations given the current lack of strong evidence of beyond the Standard Model physics at the Large Hadron Collider LHC Run 2 and other terrestrial colliders. The ability for the LHC and near future terrestrial colliders to discover new phenomena is limited by the energy scales that they can reach, while on the other hand cosmological observables have the potential to probe energies well beyond these experiments. Imprints and

© Springer International Publishing AG, part of Springer Nature 2018
N. D. Barrie, *Cosmological Implications of Quantum Anomalies*,
Springer Theses, https://doi.org/10.1007/978-3-319-94715-0_1

remnants of the dynamics of primordial high energy scale periods can potentially be seen in cosmological observables. An example of this is the Cosmic Microwave Background, which has illuminated many of the cosmological properties of our universe that we know today. Through considering a combination of observables from terrestrial collider searches and cosmological observables it may be possible to piece together the answers to many of the open questions of our universe.

This thesis will be structured, chronologically, as follows. In the remainder of this chapter we will briefly describe the Standard Models of Particle Physics and Cosmology before outlining the major questions facing these paradigms that will be addressed in this work, and the properties of anomalous symmetries in the Standard Model of Particle Physics. In Chap. 2, we will examine the inflationary epoch and a possible mechanism for inflation which also solves the hierarchy problem [1]. Chapters 3 and 4 present two possible models to explain the origin of the matter-antimatter asymmetry, the first acting during the inflationary epoch involving the introduction of a new gauge boson to the Standard Model [2, 3], and the other is a mechanism driven by the inflaton during reheating utilising the Ratchet Mechanism [4]. In Chap. 5, a novel way to utilise gravitational waves to illuminate the properties of the illusive Cosmic Neutrino Background will be considered [5]. Finally, in Chap. 6, we conclude with a discussion of the implications of each of these works and possible future paths for exploration. The focus of this thesis is on the cosmological implications of particle physics phenomena, but throughout my candidature we also conducted research into LHC Phenomenology that will not be discussed here [6–8].

1.1   The Standard Model of Particle Physics

The construction of the Standard Model of Particle Physics (SM) is one of the greatest achievements in modern science, the culmination of decades of theoretical and experimental endeavour by scientists from around the world with the goal of understanding the fundamental building blocks of nature; some of the foundational works and a review are listed [9–40]. The SM describes all of the known fundamental particles in nature and the interactions between them [41], apart from gravity which is still without a consistent quantum description. This includes all the fermions, comprising of quarks and leptons, and the gauge bosons which are the force carrying particles, as depicted in Fig. 1.1. The SM is now complete, with the last piece of the puzzle being recently discovered, the Higgs boson. The Higgs particle was theoretically predicted decades ago to explain the origin of the masses of the SM particles [42–45], and was discovered at the Large Hadron Collider (LHC) [46] by the ATLAS [47, 48] and CMS [49, 50] collaborations.

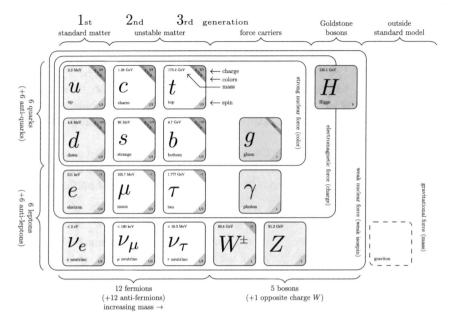

Fig. 1.1 The SM particle content, including masses, charges, and spins of each of the fermions and bosons

1.1.1 Formulation and Structure of the Standard Model

The SM is a mathematical formulation that describes all the known fundamental particles of our universe, and their interactions which are mediated by the known forces—the strong, weak, and electromagnetic forces. It combines the half-integer spin fermions, including the leptons and quarks, and the integer spin gauge bosons, consisting of the photon, weak bosons, gluons, and Higgs particle. These are each described by fields, which are mathematical objects defined at every point in space-time. It is built on the ideas of Quantum Field Theory, and the structure is fundamentally rooted in the ideas of symmetries and gauge invariance [51–58].

The SM is defined by the following direct product of groups $SU(3)_C \times SU(2)_L \times U(1)_Y$, with each group representing a gauge symmetry. The structure of the group determines the number and properties of the force carrying gauge bosons that couple to the associated charges. The SM is expressed in the form of a Lagrangian with all terms invariant under the gauge transformations described by these groups, which are local, spacetime dependent transformations. If the action is invariant under a given symmetry, then the physics derived from it will not change under such a transformation and the symmetry will have a corresponding conserved quantity. The SM gauge groups represent the following gauge bosons and forces,

- $SU(3)_C$: defines the strong force which is mediated by the gluons. These couple to fields that carry colour charge, which include the quarks and the gluons—known as Quantum Chromodynamics (QCD).
- $SU(2)_L$: the W^1, W^2 and W^3 bosons mediate the weak force, with a linear combination of W^1 and W^2 forming the $W^\pm$, while W^3 makes up part of the Z boson and photon along with the hypercharge boson. Apart from self-couplings, the weak bosons couple only to the left-handed fermions, and as such maximally violate the $\mathcal{C}$ and $\mathcal{P}$ symmetry.
- $U(1)_Y$: the hypercharge boson is a linear combination of the Z boson, and the photon. The strength of the coupling to a given particle is defined by the hypercharge $Y = I^3 + Q$, where I_3 is the weak isospin and Q is the electric charge of the particle.

Each gauge group has a corresponding gauge field strength, $G^a_{\mu\nu}$, $W^b_{\mu\nu}$ and $B_{\mu\nu}$, with their respective gauge fields G^a_μ, W^b_μ and B_μ, which define the dynamics of the gauge bosons. The gauge field strength tensor for a non-Abelian gauge group is given by,

$$F^{\mu\nu} = \partial^\mu A^\nu - \partial^\nu A^\mu + ig[A^\mu, A^\nu] \,. \tag{1.1}$$

The Higgs mechanism was proposed as a way for generating the masses of the SM particles, which involves spontaneous symmetry breaking [42–45, 59–62]. For an arbitrary scalar field its potential may contain a minimum at which the field value is non-zero, if the scalar was to rest in this minima this would be known as a vacuum expectation value. This is made more interesting by the possibility that there are many degenerate such minima that are related by a continuous symmetry, which is broken by the scalar taking a vacuum expectation value. This is what the Higgs field undergoes in the Higgs mechanism. Applying the Higgs mechanism to the $SU(2)_L \times U(1)_Y$ portion of the SM leads to spontaneous symmetry breaking, that is, $SU(2)_L \times U(1)_Y$ is broken to a single $U(1)$ gauge group, which can be identified with the photon of electromagnetism. Given that the $U(1)$ symmetry is unbroken it represents a massless boson, the photon, and indicates that the other three bosons associated with the degrees of freedom of the original gauge group structure must now be massive. The photon is associated with generators from both groups of the unbroken electroweak part of the original SM gauge group structure, which contain linear combinations of the photon and Z boson generators that are resolved after spontaneous symmetry breaking. This illustrates the unification of the weak and electromagnetic forces. In the cosmological setting, this spontaneous symmetry breaking is related to the Electroweak Phase Transition (EWPT). Once the temperature gets below a certain critical point, the Higgs boson will take a non-zero vacuum expectation value, triggering spontaneous symmetry breaking in the SM; this occurred when the temperature of the universe was about 100 GeV [63–66]. Above the EWPT, the weak gauge bosons are massless and the SM will be described by the unbroken gauge group structure.

The fermions are described in singlet or doublet representations depending on whether they have a right or left chirality, respectively. Each generation of lepton

Table 1.1 The representations of the fermions, in reference to the SM gauge groups, including corresponding baryon and lepton numbers

Fermions	$SU(3)_C$	$SU(2)_L$	$U(1)_Y$	B	L
$Q_L^i = \begin{pmatrix} u \\ d \end{pmatrix}_L^i$	3	2	$\frac{1}{6}$	$\frac{1}{3}$	0
u_R^i	3	1	$\frac{2}{3}$	$\frac{1}{3}$	0
d_R^i	3	1	$-\frac{1}{3}$	$\frac{1}{3}$	0
$L^i = \begin{pmatrix} \nu \\ e \end{pmatrix}_L^i$	1	2	$-\frac{1}{2}$	0	1
e_R^i	1	1	-1	0	1

and quark is represented by spinors, as defined by the Dirac equation. In Table 1.1, it is shown how the representation of each particle couples to each of the SM gauge groups $SU(3)_C$, $SU(2)_L$ and $U(1)_Y$.

The right-handed fermions are singlet fermions and do not couple to the weak bosons, unlike the left-handed fermion doublets. This defines the $SU(2)_L$ gauge symmetry as a chiral group. Only the quarks and gluons are charged under the $SU(3)_C$ group, with a triplet representation which corresponds to the three possible colours states. The $U(1)_Y$ gauge group of the SM describes a combination of the electromagnetic and weak Z gauge bosons. The coupling of this group to the fermions is defined as the hypercharge which is given by $Y = I^3 + Q$, where $I^3 = \frac{1}{2}\tau^3$ is known as the weak isospin and is the third generator of the $SU(2)$ gauge group, giving the connection to the Z boson. The hypercharge of each fermion is given in Table 1.1.

The SM is defined by a Lagrangian density, which contains all the renormalisable terms that represent the possible field interactions, kinetic and mass terms. The Lagrangian describes the dynamics and evolution of each of the fundamental particles. The allowed terms in the Lagrangian are determined by the symmetries of the model, and how the fields transform under them, with the requirement that the action must be left unchanged under any transformations associated with the SM gauge symmetries, or assumed global symmetries. Imposing renormalisability requires the dimension of the operators in each term of the Lagrangian density to be ≤ 4, such that the coefficients are dimensionless or have positive mass dimension. In any case, higher dimensional terms are generally suppressed at low energies by the mass scale of the coefficients. The SM Lagrangian density is given by,

$$\mathcal{L}_{SM} = \mathcal{L}_{Gauge} + \mathcal{L}_{Dirac} + \mathcal{L}_{Scalar} + \mathcal{L}_{Yukawa} , \tag{1.2}$$

where

$$\mathcal{L}_{Gauge} = -\frac{1}{4}G_{\mu\nu}^a G^{a\mu\nu} - \frac{1}{4}W_{\mu\nu}^b W^{b\mu\nu} - \frac{1}{4}B_{\mu\nu}B^{\mu\nu} , \tag{1.3}$$

describes the gauge fields and their related bosons,

$$\mathcal{L}_{Dirac} = i\bar{\psi} D_\mu \psi \, , \tag{1.4}$$

describes the free fermions, where ψ is any of the fermion representations within the theory,

$$\mathcal{L}_{Scalar} = (D_\mu \phi)^\dagger (D^\mu \phi) + m^2 \phi^\dagger \phi - \frac{\lambda}{2}(\phi^\dagger \phi)^2 \, , \tag{1.5}$$

describes the Higgs kinetic term and potential,

$$\mathcal{L}_{Yukawa} = -y_{q_R} \bar{Q}_L \phi q_R - y_{e_R} \bar{L} \phi e_R + h.c. \tag{1.6}$$

describes each of the fermion interactions between the left-handed particles and their right-handed counterparts, via the Higgs boson. These terms generate the masses of the fermions, while the first term in Eq. (1.5) generates those for the $W^\pm$ and Z bosons, once the Higgs particle takes a non-zero vacuum expectation value. The covariant derivatives D_μ incorporate the interactions between the gauge fields and the SM fields.

It is possible the SM may not extend to high energies and is instead a low energy Effective Field Theory [67], which would indicate why it is unable to explain early universe phenomena. Meaning that it is valid below some energy scale, but breaks down above it, with new physics required; analogously to the divide between classical and quantum mechanics.

1.1.2 Symmetries in the Standard Model

The SM is built on the principles of gauge invariance through the description of the force carrying particles as vector bosons. This identification naturally leads to the requirement of gauge invariance associated with these vector fields, because without it the theory would be non-unitary and Lorentz violating, with the associated gauge bosons potentially having tachyonic degrees of freedom. These things would lead to an inconsistent and non-predictive theory. This means that the gauge symmetries are redundant, in that, the theory would not exist without their imposition. As such, these symmetries dictate what the allowed interactions are within the theory, or rather what terms can be contained within the Lagrangian. Gauge symmetries must also be protected from quantum anomalies—breakdowns in gauge invariance due to radiative corrections. They must be conserved at all orders of perturbation. Although, this is not necessarily required for global symmetries.

Classical global symmetries are characteristic symmetries of a theory at the tree level, and can be retained at all orders of perturbation. They can be present due to all the interactions allowed by the gauge symmetries respecting the symmetry acci-dentally, or they can be explicitly imposed on the theory. Each global symmetry in a theory has a corresponding charge which is conserved by the classical interactions. Although, global symmetries do not have to necessarily hold once quantum corrections are included. Symmetries broken by radiative corrections are known as anomalous symmetries. Gauge symmetries cannot be broken in this way because that would lead to a breakdown in the consistency of the theory through violations of gauge invariance.

Renormalisability requires that the mass dimension of each operator adds up to four or less, meaning that all coefficients must be dimensionless or have positive mass dimension. An accidental symmetry of the SM is one which is only present due to the requirement that the Lagrangian terms are renormalisable. Imposing this condition causes the removal of higher order terms from the Lagrangian that would have otherwise broken certain symmetries, subsequently retaining them as symmetries. Some examples of accidental symmetries of the SM Lagrangian are those associated with the global baryon number, and generational lepton numbers.

Global Baryon and Lepton Numbers

Baryon number is an accidental global symmetry of the SM. In the tree level SM, there are six possible $U(1)$ quark symmetries that correspond to each quark flavour, but are broken once the Cabbibo-Kobayashi-Maskawa (CKM) interactions are considered. This leads to there being only a single global $U(1)$ symmetry associated with the quarks, the total quark number, which can be identified with the baryon number symmetry of the SM. Each quark is assigned a baryon number charge of $\frac{1}{3}$ under this symmetry, as given in Table 1.1. The corresponding transformation is given by the following unitary transformations $q \rightarrow e^{i\frac{1}{3}\beta}q$, where β is a real number and q represents a quark field.

The generational lepton numbers are also accidental global symmetries of the SM. These are $U(1)$ symmetries given by the unitary transformations $l_i \rightarrow e^{i\alpha_i}l_i$, where α_i is a real number and i denotes the lepton generation. Although, the discovery of neutrino oscillations proved that these symmetries are violated. Instead the SM contains a single $U(1)$ associated with the total lepton number, as the oscillations still leave the total lepton number conserved [56]. Each of the leptons is assigned a lepton number equal to 1, as given in Table 1.1.

The corresponding Noether currents for the continuous global baryonic and leptonic symmetries are given by [68],

$$j_B^\mu \propto \bar{u}_i\gamma^\mu u^i + \bar{d}_i\gamma^\mu d^i \text{ and } j_L^\mu \propto \bar{l}_i\gamma^\mu l^i + \bar{\nu}_i\gamma^\mu \nu^i , \qquad (1.7)$$

where these lead to conservation of B and L at the tree level, $\partial_\mu j^\mu = 0$, and i denotes the fermion generation.

The $U(1)_B$ and $U(1)_L$ symmetries are present in the SM due to the imposition of renormalisability. There are non-renormalisable operators which can be written down that are consistent with the SM gauge symmetries, but explicitly break the baryon and lepton number symmetries. The lowest dimension operators that are examples of these are $\frac{QQQL}{\Lambda_B^2}$ and $\frac{LLHH}{\Lambda_L}$, respectively. These terms are strongly suppressed by their dimensional coefficients. For example, the first term gives a path for proton decay which, from experiment, has a decay time of $\tau_p > 10^{32-34}$ years, implying $\Lambda_B > 10^{15}$ GeV [69]. The high energy scales associated with these terms, Λ_B and Λ_L, leads to the suppression of these interactions at lower energies, but they could be indicative of new physics at those scales.

In this thesis, we are interested in symmetry violation by quantum anomalies, that is, induced by radiative corrections.

1.1.3   Radiative Corrections and Quantum Anomaly Cancellation

Anomalous symmetries are symmetries of the tree level SM that are violated once quantum corrections are included [70–72]. The SM requires radiative corrections of the tree level interactions so that precise predictions can be made, and so all the possible interactions can be determined and be taken into account. This can be done by perturbatively expanding the tree level vertices associated with interactions, in Feynman diagrams, into loops of a higher and higher order, where each subsequent expansion is suppressed relative to the previous one. These loop calculations give the quantum corrections to the corresponding interactions, and may even introduce new interactions that are not present at the tree level. Some examples of loop diagrams are presented in Fig. 1.2, where in the cases presented, there is no tree level vertex with the same initial and final states, and hence such interactions are only possible with radiative corrections. These diagrams are also special, in that they may correspond to anomalous contributions to the associated baryon number symmetry, depending on the fermionic content of the theory.

Quantum anomalies can be induced for global symmetries, violating the conserved charges to which they correspond. The anomalies of a given symmetry are produced by radiative corrections, the magnitude of which are found by considering what particles couple to the vertices of the expanded loop. This is determined by taking the trace of the vertex generators, doing so considers all the possible fermions that could traverse the loop, with the resultant coefficient defining the magnitude. If zero, there is no anomaly present and the associated symmetries are preserved at that order of expansion. If non-zero, this gives you a correction to the Noether current of the corresponding global symmetry, which causes it to no longer be conserved, $\partial_\mu j^\mu \neq 0$. An example of this is shown in Fig. 1.2, for the global baryon number current in the SM.

The gauge fields that couple to each vertex determine the possible fermions that can traverse the loop. For example, only left-handed fermions couple to the $SU(2)$ gauge field, so only q_L and l_L make up the possible constituents of the fermionic current in the corresponding loop corrections. There are similar constraints for $SU(3)$, that is, only the quarks Q_L, u_R and d_R are considered—while all SM fermions couple to the hypercharge gauge group $U(1)_Y$. Refer to Appendix A, for calculations of

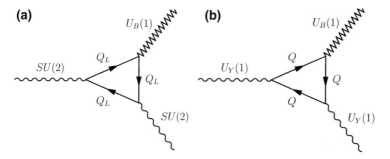

Fig. 1.2 The 1-loop anomalous radiative corrections involving a gauge boson associated with the $U(1)_B$ symmetry, **a** $SU(2)^2U(1)_B$, where the loop contains only left-handed quarks, and **b** $U(1)_Y^2U(1)_B$ where the loop contains only quarks

each of the 1-loop contributions for the baryon number and lepton number currents, as well as in the case that they are promoted to gauge symmetries.

Examples of anomalous symmetries in the SM are the global baryon and lepton number symmetries. In calculating the radiative corrections, the inclusion of a baryon number current means that only the quarks are considered in the fermionic loops. If one was to consider these as gauged symmetries, for example a gauged baryon number, then upon calculating the first order loop diagrams non-zero anomalies are found to be present for the $SU(2)^2U(1)_B$ and $U(1)_Y^2U(1)_B$ interactions. These vertex expansions are shown in Fig. 1.2. This means that the baryon number symmetry has the following non-zero triangle anomalies:

For $SU(2)^2U(1)_B$, where only left-handed quarks are considered,

$$A_2(SU(2)^2U(1)_B) = Tr[\tau^a\tau^b B] = \frac{n_g}{2} = \frac{3}{2} . \tag{1.8}$$

For $U(1)_Y^2U(1)_B$, where all quarks contribute,

$$A_3(U(1)_Y^2U(1)_B) = Tr[YYB] = -\frac{n_g}{2} = -\frac{3}{2} . \tag{1.9}$$

where n_g is the number of quark generations, which is taken to be three.

It is found that the global baryon and lepton number symmetries have the same quantum anomalies when right-handed neutrinos are included in the SM, but with $U(1)_L$ replacing $U(1)_B$ in Eqs. (1.8) and (1.9). This indicates that $B - L$ is still a conserved global symmetry of the SM after quantum corrections are included.

The anomalous currents associated with the $U(1)_B$ and $U(1)_L$ currents source B and L violating processes, namely electroweak instanton transitions and sphaleron processes. These shall be discussed in Chap. 3, in the context of how this violation of B and L is utilised in Electroweak Baryogenesis, and how it can also be responsible for the washout of asymmetries generated early in the universes evolution.

The anomaly calculations of each of the quantum corrections for the $U(1)_B$ and $U(1)_L$ symmetries are given in Appendix A. These anomalies provide avenues for baryon and lepton number violation, as they are not strictly conserved quantities at the quantum level. If the $U(1)_B$ and $U(1)_L$ symmetries are introduced as local gauge symmetries these anomalies become gauge anomalies. This is a major issue for the validity of the theory and they must be cancelled, the methods of which we shall now discuss.

Gauge Anomaly Cancellation and the Green-Schwarz Mechanism

Quantum anomalies associated with a gauge symmetry must be cancelled to maintain the consistency of the theory. The reasons for this are two fold, firstly, they lead to a loss of unitarity of the S matrix and violation of Lorentz invariance, and secondly, they cause a breakdown in renormalisability. Each of which is required for a gauge invariant theory [73]. Such gauge anomalies can be cancelled in various ways through the inclusion of new degrees of freedom. Two examples of ways in which this can be done are the addition of new fermions, and the Green-Schwarz mechanism [74].

The inclusion of extra fermions can be used to cancel the anomalies induced by radiative corrections, as their inclusion leads to additional terms being present in the trace of the vertex generators. The quantum numbers of these new fermions can then be chosen such that the anomalies disappear. These can create new problems as they could mediate interactions which have not been observed at terrestrial colliders, or could form charged stable fields that could be detrimental to cosmological observables. Experimental investigations provide restrictions on the possible masses and couplings of any introduced particles, and the requirement of complete cancellation of the anomalies means that they can only carry certain quantum numbers.

The Green-Schwarz mechanism was first proposed in the context of string theory, due to the appearance of unwanted gauge anomalies in such theories. To resolve this issue Green and Schwarz developed a method of restoring gauge invariance via the addition of new terms to the effective action [74]. This leads to an effective field theory in which a new degree of freedom has been introduced to the theory. These new terms remove the anomalies by cancelling the non-invariance of the fermion measure in the action, leading to a gauge invariant theory that effectively describes the dynamics of the full theory at energies lower than a characteristic scale.

As an example of how this mechanism works, consider the inclusion of a new $U(1)_X$ symmetry to the SM, which has gauge anomalies. In this case, the Green-Schwarz terms added to the effective action contribute a new longitudinal degree of freedom to the introduced gauge field X_μ, which corresponds to the gauge field acquiring a mass. In doing so, we obtain an effective field theory, meaning that it can only sufficiently describe the interactions of the theory well below some new physics scale, which we shall denote f_X. This parametrises our ignorance of the full theory, which one could imagine contains heavy fermions that cancel the anomaly at scales well above f_X. For example, the full theory may involve the inclusion of leptoquarks at high energies, which can cancel the gauge anomalies present in the theory [75–77].

The new counter-terms added to the effective action are included in the Lagrangian density as follows,

$$\mathcal{L}_{counter} = -\frac{g_X^2}{16\pi^2}\mathcal{A}\theta X_{\mu\nu}\tilde{X}^{\mu\nu} , \qquad (1.10)$$

where g_X is the coupling for the gauge field related to the anomaly, $X^{\mu\nu}$ is the field strength tensor, and $\mathcal{A}$ is the associated anomaly coefficient. The θ component is pure gauge and is the longitudinal degree of freedom of the new gauge field X_μ. It is found that the tree level SM Lagrangian appears to not be gauge invariant once these counter-terms are added, although their variance cancels out against the gauge variation of the functional measure of fermion fields within the path integral quantization framework, leading to a gauge invariant theory. This term is reminiscent of the term for a massive gauge boson in the Stuekelberg formalism [78].

Examples of mixed gauge anomalies are those that result from promoting the global baryon number symmetry to a gauge symmetry, for which multiple anomalies are present. The following counter-terms must be included,

$$\mathcal{L}_{counter} = -\frac{3g_2^2}{32\pi^2}\theta(x)W_{\mu\nu}\tilde{W}^{\mu\nu} + \frac{3g_1^2}{32\pi^2}\theta(x)B_{\mu\nu}\tilde{B}^{\mu\nu} . \qquad (1.11)$$

In this thesis, we shall be considering the cosmological implications of quantum anomalies such as these, and whether they can be utilised to explain various mysteries of the SM and Standard Model of Cosmology (SMC).

1.2 The Standard Model of Cosmology and the Evolution of the Universe

The Lambda Cold Dark Matter (ΛCDM) model is the current best model we have for describing the evolution of the universe from its beginning to the present day, and is considered the SMC [79–82]. The SMC incorporates all known luminous matter, cold dark matter, and a cosmological constant to describe the dark energy density. The present day total energy content of the universe is split between these three main components—ordinary matter ($\sim$5%), dark matter ($\sim$27%), and dark energy ($\sim$68%). Observations of the Cosmic Microwave Background (CMB) in combination with other observational surveys provide data on the cosmological parameters of this model.

The measurement of the CMB has provided us a window into the early history of the universe [83–93]. Much of what we know about the universe comes from analysing the CMB and the temperature fluctuations that can be found within it. Knowledge of the properties of the universe prior to the formation of the CMB is difficult to glean due to the opaque nature of the primordial plasma prior to recombination. Despite this, the CMB can carry information from earlier times, such as inflation, allowing constraints to be placed on the properties of these prior epochs.

Fig. 1.3 The Cosmic Microwave Background as measured by the PLANCK satellite [94]

In the case of inflation, this can be done through the spectral index and tensor to scalar ratio of the CMB perturbations for which inflation makes predictions (Fig. 1.3).

The CMB is found to be highly isotropic and homogeneous on large scales, and as such, in the SMC the universe is considered to be static in conformal coordinates. Meaning that the universe evolves homogeneously and isotropically defined by a global scale factor $a(t)$. The gravitational metric associated with these evolutionary properties is known as the flat Friedmann-Robertson-Walker (FRW) universe, which is strongly supported by observations of the CMB. This is the basis of our understanding of the evolution of the universe, and shall be discussed below.

1.2.1 The Friedmann-Robertson-Walker Metric

One of the simplest models used to describe the evolution of the universe is the set of FRW cosmological models, which are a particular set of solutions of the Einstein equation that describe isotropic and homogeneous universes [95]. Isotropy means that they are spherically symmetric around a point in space and hence have no preferred direction, while homogeneity means that every point in space is equivalent, and hence there is no preferred position. This is supported by the remarkable uniformity of the CMB, observed by the WMAP [90] and PLANCK satellites [94]. Exact homogeneity would imply that there must also be no boundary or centre of the universe. For simplicity the universe is modelled as an isotropic and homogeneous spacetime.

This class of models includes three possible spatial geometries—closed, flat and open; each with constant curvature. A closed geometry is analogous to a sphere, which has constant positive curvature, flat to a plane, which has zero curvature, and open to a hyperbolic plane, which has constant negative curvature. For a general FRW universe, the metric in polar coordinates is given by,

$$ds^2 = dt^2 - a^2(t) \left(\frac{dr^2}{1 - kr^2} + r^2(d\theta^2 + \sin^2\theta d\phi^2) \right), \quad (1.12)$$

where k defines the spatial curvature of the metric, and takes the values $k = 1, 0$, or -1 which correspond to closed, flat and open spaces, respectively. The scale factor $a(t)$ embodies the rate of expansion or contraction of the universe.

From observation, we will be assuming that the universe is a simple flat FRW universe as the SMC considers, which has the following metric in Cartesian coordinates,

$$ds^2 = dt^2 - a^2(t)(dx^2 + dy^2 + dz^2), \quad (1.13)$$

or in conformal coordinates, which is defined by the following transformation of the time coordinate $dt = a(\tau)d\tau$,

$$ds^2 = a^2(t)(d\tau^2 - dx^2 + dy^2 + dz^2), \quad (1.14)$$

and hence the metric can now be written as $g_{\mu\nu} = a(\tau)^2 \eta_{\mu\nu}$, which is conformally flat. Therefore, this flat FRW metric is a suitable model for the spacetime of our universe and provides a simple framework in which to describe its evolution.

From the FRW metric can be derived the well-known Friedmann Equations, utilising the Einstein equation. They relate the nature of the expansion of the universe to the energy densities and curvature that is present, as described in Table 1.2. The Friedmann equations are as follows,

$$H^2 = \left(\frac{\dot{a}}{a} \right)^2 = \frac{8\pi\rho}{3M_p^2} + \frac{\Lambda}{3} + \frac{k}{a^2}, \quad (1.15)$$

$$\frac{\ddot{a}}{a} = -\frac{4\pi}{3M_p^2}(\rho + 3p) + \frac{\Lambda}{3}. \quad (1.16)$$

where ρ and p are the mass density and pressure of the perfect fluid, respectively, k is the Gaussian curvature defined above, M_p is the Planck mass, H is the Hubble rate [96], and Λ is the cosmological constant.

The evolution of the universe is defined by the properties of the scale factor, which is determined by the form of energy that is dominating. This is illustrated in the time evolution of the scale factors shown in Table 1.2, which are epoch dependent. In this table we use the dimensionless parameter w, known as the equation of state, to define each epoch. This parameter is related to the energy density by,

$$\rho \propto a^{-3(1+w)}. \quad (1.17)$$

Table 1.2 Scale factors corresponding to each form of energy in a flat FRW universe

Type of energy	w	$\rho(a)$	$a(t)$	$a(\tau)$
Matter dominated	0	$\propto a^{-3}$	$\propto t^{2/3}$	$\propto \tau^2$
Radiation dominated	$\frac{1}{3}$	$\propto a^{-4}$	$\propto t^{1/2}$	$\propto \tau$
Vacuum dominated	-1	Constant	$\propto e^{Ht}$	$\propto \frac{-1}{H\tau}$

1.2.2 The History of the Universe

The main epochs and events in the history of the universe, which we will be interested in, are as follows:

Inflation: Shortly after the Big Bang, or pre-inflationary state, the universe underwent a rapid period of expansion within a short period of time. This expansion is able to explain the homogeneity and flatness observed in the CMB, as well as the temperature fluctuations found in it. The exact nature of the mechanism leading to inflation is yet unknown, but it is thought to be caused by a particle known as the inflaton which dominates the energy density of the early universe. The induced period of inflation is an approximate vacuum dominated era, during which the Hubble parameter is approximately constant and the scale factor is accelerating.

Reheating: After inflation the universe begins the transition to a radiation dominated epoch through a period known as reheating. The characteristics of the reheating epoch and the temperature of the resultant radiation dominated universe is determined by the properties of the inflationary mechanism and its interactions with the SM or mediator particles. The reheating temperature can be as high as $\sim 10^{15}$ GeV, for a high inflationary scale, and as low as ~ 1 MeV to not conflict with predictions from Big Bang Nucleosynthesis (BBN). The exact properties of reheating are inherently complicated and are still not fully understood.

Baryogenesis: The universe has been observed to have an asymmetry between matter and antimatter, with minimal antimatter found to be present. The dynamics which lead to the origin of this asymmetry are generally assumed to occur at or above the electroweak scale, $T \sim 100$ GeV. This is highly dependent on the production mechanism, and could potentially happen as late as $T \sim 1$ MeV; once again, to not conflict with BBN.

Dark Matter Freeze-out: It is thought that the dark matter content of our universe was produced sometime during the radiation epoch. After being produced it would have thermally decoupled from the hot SM plasma once the temperature of it became low enough to suppress thermal production; assuming the coupling of the dark matter to the thermal plasma is small. As the interactions between the SM particles and dark matter are assumed to be fairly weak, this freeze out can occur fairly early on, meaning that the density is only affected by dilution due to the spatial expansion if the dark matter is stable relative to the age of the universe.

Electroweak Phase Transition: At $T \sim 100$ GeV the Higgs potential becomes such that the Higgs field takes a non-zero vacuum expectation value, leading to spontaneous symmetry breaking. As the temperature lowers, thermal fluctuations will cause certain regions to take a vacuum expectation value before others, leading to the formation of bubbles of true and false vacuum. This phenomenon has been considered as the origin of the observed baryon asymmetry, with this mechanism being known as Electroweak Baryogenesis.

Neutrino Decoupling: As the neutrinos only couple via the weak interaction to the SM, they decouple from the plasma relatively early in the history of the universe, $T \sim 1$ MeV. The decoupled neutrinos make up the contents of the Cosmic Neutrino Background (CνB), which like the CMB, permeates the universe today.

Electron Positron Annihilation: At $T \sim 500$ keV the thermal production of $e^+ - e^-$ pairs freezes out, and the annihilation of these pairs, in the plasma, into photons will leave only the asymmetric part remaining. This annihilation increases the photon energy density of the universe.

Big Bang Nucleosynthesis: This is the period in which the nuclei observed in the universe today have their origin, through a series of well understood nuclear fusion reactions. This is well modelled by current nuclear physics models, and as such is very sensitive to new physics [97–107].

Recombination and Photon Decoupling: At a certain point in the evolution of the universe its temperature will be low enough that the electrons, and the nuclei generated in BBN, will begin to form atoms such as neutral hydrogen. As more nuclei form the universe becomes transparent to photons, which allows them to decouple from the plasma. This leads to the formation of the CMB, at the end of recombination, which we observe today.

Dark Energy-Matter Equality: In more recent universal history, the dark energy density became larger than the matter energy density of the universe. This vacuum domination produces an accelerated expansion, which may be associated with a cosmological constant, and is the current state of the universe (Table 1.3).

1.3 Mysteries to be Solved

Despite the successes of the SM and SMC in describing the nature and evolution of the universe, there are still many unsolved mysteries. In this thesis, we hope to illuminate some of the following open problems.

1.3.1 The Inflationary Mechanism

The existence of an inflationary epoch was first postulated to solve unexplained observed phenomena and theoretical issues in cosmology [109–121]; namely, the flatness, horizon, and monopole problems [81, 122]. Since this time, significant

Table 1.3 The key events in cosmological history [108]

Event	Time t	Redshift z	Temperature T
Inflation	10^{-34} s (?)	–	–
Baryogenesis	?	?	?
Dark matter freeze-out	?	?	?
Electroweak phase transition	20 ps	10^{15}	100 GeV
QCD phase transition	20 µs	10^{12}	150 MeV
Neutrino decoupling	1 s	$6 \cdot 10^9$	1 MeV
Electron positron annihilation	6 s	$2 \cdot 10^9$	500 keV
Big Bang Nucleosynthesis	3 min	$4 \cdot 10^8$	100 keV
Matter-radiation equality	60 kyr	3400	0.75 eV
Recombination	260–380 kyr	1100–1400	0.26–0.33 eV
Photon decoupling	380 kyr	1000–1200	0.23–0.28 eV
Re-ionisation	100–400 Myr	11–30	2.4–7.0 meV
Dark energy-matter equality	9 Gyr	0.4	0.33 meV
Present	13.8 Gyr	0	0.24 meV

amounts of evidence supporting a period of cosmic inflation has been obtained from a number of astrophysical observations [92], leading to it being a generally accepted part of the early universe evolution.

The flatness problem stems from the observed flatness of the universe, known to a very high precision, which appears to be fine-tuned. Rather than flatness being a special initial condition, inflation provides a natural way for such a lack of curvature to be the observed state of the universe. If one was to consider an initially curved spacetime, it would be approximately flat at very small scales. The rapid expansion of the universe produced by inflation will make these locally flat regions much larger, such that on large scales the universe will begin to appear flat. That is, the rapid spatial expansion quickly dilutes spatial curvature, pushing it towards zero. Hence the observable universe appearing flat is a natural consequence of inflation, if enough e-folds of expansion have occurred, resolving the problem.

A major mystery that inflation provides a solution for is the so-called horizon problem. This is the observation that the universe is very homogeneous and appears to have been in thermal equilibrium, although if one is to trace back the evolution of the universe to the Big Bang it is found that all of today's causally disconnected regions would not have been in causal contact. This is because the universe has existed only for a finite period of time, so information can only have travelled a finite distance. This raises the question of why these causally disconnected regions are so well correlated. An inflationary epoch can provide the answer to this question

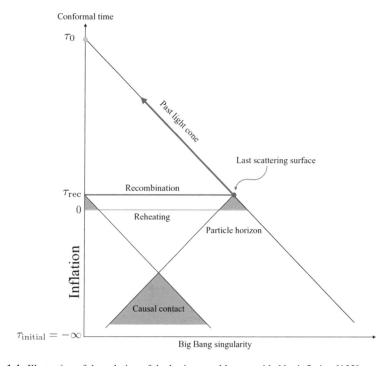

Fig. 1.4 Illustration of the solution of the horizon problem provided by inflation [122]

by allowing the universe to have been in thermal equilibrium at very early times, prior to the onset of inflation. If the very early universe, just after the big bang, was in thermal equilibrium and then an inflationary epoch began, the resultant universe would be broken into many causally disconnected regions, after enough e-foldings, explaining what we observe today. This idea is depicted in Fig. 1.4.

A generic prediction of many Grand Unified Theories is the production of topological defects, such as monopoles, at phase transitions which occur as the universe cools from the very high temperatures at the end of reheating. The dynamics of the early universe would lead to the thermal production of a high density of monopoles. Unfortunately, it is found that the energy density of these monopoles in the universe would be too large to be consistent with current observations. This is known as the monopole problem [123–125]. Now if one is to postulate an inflationary epoch, this would lead to the significant dilution of the density of these relics, removing them as a potential issue, assuming the reheating temperature is low enough to prevent their overproduction after inflation.

Another attractive property of the inflationary scenario is that it can generate the density perturbations observed via the inhomogeneities of the CMB. These density perturbations provided the seeds for large scale structure formation, and hence led to the formation of the galaxies we see today. As inflation proceeds, quantum fluctuations of the inflaton field, the scalar field which induces the inflationary epoch, are enlarged by the rapid spatial expansion, producing the temperature anisotropies we see in the CMB [126].

Once inflation ends the initial energy densities of matter have been diluted to negligible quantities, so the universe must be reheated. This is also taken into account in the inflationary mechanism, through the decay of the inflaton after inflation ends; this period is known as the reheating epoch. The product of the reheating epoch, is a radiation dominated universe with a thermalised plasma with characteristic temperature T_{rh}, after which the standard big bang cosmology proceeds.

Unfortunately, despite the many successes of the inflationary paradigm at explaining cosmological evolution, the exact mechanism for inflation is still not known. It is usually assumed that this epoch is the result of a scalar field, named the inflaton, which slowly rolls towards the minimum of its potential [122, 127–132]. The potential energy of this scalar field dominates the energy density of the universe, as long as the inflaton only rolls slowly in its potential. This almost constant potential energy, dominating at early times, leads to an effective vacuum energy dominated universe; a de Sitter spacetime. Such a vacuum energy dominated universe is characterised by exponential expansion, as discussed above. It should be noted that cosmological models other than inflation have been proposed to solve the issues described above, such as String gas and bounce cosmologies, with varying success [133–135].

Many inflaton candidates have been proposed since inflation was first postulated, with most involving the introduction of a scalar(s), but as of yet none have been experimentally verified [115, 126, 136–157]. Due to this, many attempts to constrain and test inflationary mechanisms have been considered [158–166]. One reason this is difficult is that with the current experimental sensitivity to inflationary observables, many models give degenerate solutions and hence are indistinguishable. This has led to recent work into identifying classes of inflationary models [167–170]. Another issue is that the inflaton itself is unlikely to be produced terrestrially due to it generically having a very high mass, well beyond the range of any current or future collider experiments. In Chap. 2, we will consider a new class of scale invariant inflationary scenarios which may have interesting implications for particle physics.

1.3.2 The Origin of the Matter-Antimatter Asymmetry

One of the major questions in modern physics is how the observed matter-antimatter asymmetry of the universe developed [171–184]. Antimatter was first predicted to be on equal footing with matter [185], but observational and experimental results suggest that this is incorrect. Terrestrial experimental investigations confirmed this in decays of K^0 ($d\bar{s}$) [186, 187], D^0 ($c\bar{u}$) [188] and B^0 ($s\bar{b}$) [189, 190] mesons, which have provided evidence for $\mathcal{C}$ and $\mathcal{CP}$ violation. As an asymmetry is observed, astronomical observations provide strong contradictory evidence to the equivalence of matter and antimatter [90, 94, 191, 192].

Observations indicate that the visible universe is dominated by matter and not antimatter. This baryon number density is determined by analysing the Baryon Acoustic

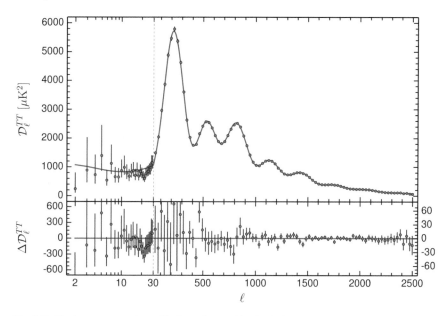

Fig. 1.5 The baryon acoustic oscillations observed in the CMB power spectrum, as measured by the PLANCK satellite [193]

Oscillations (BAO) measured from the CMB power spectrum, depicted in Fig. 1.5, and the temperature fluctuations in the CMB [90, 92, 193]. BBN theory also accurately predicts the abundances of the primordial light elements and is sensitive to the size of the baryon asymmetry; providing constraints on it [104, 194–197]. The observed value of the matter-antimatter asymmetry of our Universe is determined from the combination of these measurements, and is parametrised by the baryon to entropy density ratio,

$$\eta_B = \frac{n_B - n_{\bar{B}}}{s} \simeq \frac{n_B}{s} \simeq 8.5 \cdot 10^{-11} \, , \qquad (1.18)$$

where n_B ($n_{\bar{B}}$) is the baryon (antibaryon) number density, and s is the entropy density of the universe. It is possible that the universe may have begun with a net baryon number prior to inflation, but any baryon number density would be diluted away by the rapid expansion of inflation, hence it is assumed that it must be produced dynamically. The process of dynamical generation of the observed matter-antimatter asymmetry is known as Baryogenesis.

It has been proposed that the net baryon number of the universe is indeed zero and that many unmixed islands of matter and antimatter exist [198–200]. If separate sectors do exist it would be possible to observe annihilations at the boundaries between the regions. Electron-positron pairs involved in annihilation processes would produce high energy photons, which if in significant numbers can cause a skewing of the CMB spectrum. These processes can also heat the ambient plasma leading to an additional indirect spectral distortion. This would be observable as dilutions or perturbations in the CMB [201]. These gamma ray sources are not observed, and the CMB is found to be highly uniform, so to be consistent with observation the voids between these matter and antimatter sectors would be required to be large. However, the size of the possible voids between these sectors is also constrained by the CMB. Regions large enough to survive recombination would be observable in the CMB, but these have not been found; excluding the idea that there are sectors of matter and antimatter separated by voids within the observable universe. It is possible that we live in a matter island that is larger than the observable universe, but this is not supported experimentally [202–204].

Seeing as there is overwhelming evidence for the existence of the baryon asymmetry in the universe, we must consider how such an asymmetry could be produced.

The Sakharov Conditions

The Sakharov conditions [205], formulated by A.D. Sakharov in 1967, are the requirements for successful Baryogenesis in the early universe. The conditions are,

- **Baryon number violation** If immediately after the Big Bang there was zero net baryon number (B) and B is strictly conserved, then the net baryon number density of the universe would remain zero for all time. Therefore, B number violating processes are needed.
- C **and** CP **violation** The baryon number violating processes are required to violate the C and CP symmetries, such that either the matter or antimatter process is favoured.
- **A period of non-equilibrium** The processes that violate B, C and CP must occur in a period of non-equilibrium, so that the reverse reactions do not washout any generated asymmetry.

Any mechanism that wishes to produce a charge asymmetry in the early universe must satisfy these criteria, unless the theory violates CPT.

Baryogenesis in the Standard Model

One of the most studied mechanisms for the generation of the baryon asymmetry is Electroweak Baryogenesis. The reason for this is that the first model discovered in this class of Baryogenesis mechanisms was the SM itself, which contains all the ingredients for satisfying the Sakharov conditions [206–217]. They are satisfied in the following ways:

- **Baryon number violation** The quantum anomalies associated with the global baryon number symmetry and the electroweak gauge group indicate that it is not a strictly conserved quantity within the SM. Through these anomalies, baryon number violation can occur via non-perturbative electroweak sphaleron transitions [72], which shall be discussed further in Chap. 3.
- C **and** CP **violation** The C and P symmetries are maximally violated in the SM by the chiral nature of the weak interactions. There is also CP violation provided by a complex phase present in the CKM matrix.

Fig. 1.6 A schematic of the
Electroweak Baryogenesis
mechanism [220]

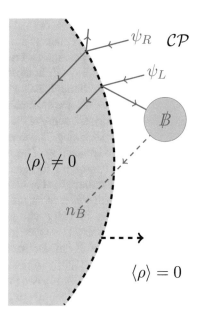

- **A period of non-equilibrium** The electroweak phase transition (EWPT) is the
  push out-of-equilibrium utilised in SM Baryogenesis [65, 218, 219]. Once the
  universe cools, thermal fluctuations in the primordial plasma can lead to regions
  falling below the EWPT earlier than others; the EWPT occurs at a characteristic
  temperature $T \sim 100$ GeV. This leads to the formation of 'bubbles', or regions, of
  broken phase, where the Higgs boson has acquired a vacuum expectation value.
  The expansion of the walls of these bubbles creates the non-equilibrium conditions
  required for Baryogenesis. This process is depicted in Fig. 1.6, where the $\mathcal{CP}$
  and B violating processes associated with the left-handed fermions lead to the
  accumulation of a net baryon number in the interior of the expanding bubble.

Although the SM satisfies the Sakharov conditions, the resulting prediction for the
baryon asymmetry generated is found to be $\eta_B^{sm} \simeq 10^{-18}$, approximately eight orders
of magnitude lower than the observed value [221]. One reason for this discrepancy is
that the strength of the $\mathcal{CP}$ violation is too small to produce a large enough asymmetry.
Also, the Higgs mass is too high, $m_h \sim 125$ GeV, for the EWPT to be strongly first
order, so the departure from equilibrium is too weak [48, 217, 222, 223]. Therefore,
the SM is not sufficient to explain the observed baryon asymmetry of the universe,
and the existence of new physics is required.

Other Proposed Scenarios

Many models have been proposed to solve the issue of the baryon asymmetry of the
universe (a selection of reviews are given here Ref. [224–235]), many of which are
simple extensions of the SM [236–239]. Some of these models also try to resolve other
unexplained physical phenomena alongside the generation of the baryon asymmetry,

such as dark matter [240–242]. As of yet most of these models remain unproven because the related phenomena are beyond the reach of current experiments.

A very attractive paradigm for the generation of the matter-antimatter asymmetry is Leptogenesis [243–249]. In these models the asymmetry is initially generated in the neutrino sector, through extensions to the SM that also explain the origin of the neutrino masses. Once the asymmetry is generated the $B + L$ violating electroweak sphalerons redistribute some of the net lepton number into the baryonic sector, producing what we see today. The interest in this form of scenario is driven by the lack of a complete understanding of the neutrino sector, which could potentially be hiding key information to solving this and other mysteries in particle physics.

Extensions that have the baryon asymmetry of the universe generated before the inflationary epoch risk having the accumulated baryon number being completely diluted. Also, the difficulty with producing the asymmetry during inflation [250–252] is the requirement that the rate of asymmetry production must be greater than the rate of dilution. If not, the generated asymmetry will be quickly diluted away. For this reason the majority of postulated mechanisms occur after reheating. In this work we will be considering two possible extensions of the SM that can explain the baryon asymmetry of the universe; one acting during inflation [2, 3] and the other during reheating [4].

1.3.3   Dark Matter

The existence of dark matter, an abundant form of non-luminous matter in the universe, was first proposed by Zwicky, in 1933, to explain the discrepancy between the expected and observed luminosity of galaxies, given their measured gravitational masses [253]. Further evidence for dark matter was provided by the observation of anomalous galactic rotation curves [254–260], which to be resolved with theoretical predictions of the rotation profile required the existence of more luminous matter than could be seen [261, 262]. Since this time the existence of dark matter has been generally accepted as many more pieces of observational evidence have been gathered; these include gravitational lensing, the BAO, and its importance for cosmological evolution and structure formation, seeing as it is a key component of the ΛCDM cosmological model. Despite this, the true identity and nature of the constituents of dark matter are not known.

One of the most researched candidates for dark matter are Weakly Interacting Massive Particles (WIMPs) [263–265]. Originally, this model was very strongly favoured because of the so-called WIMP miracle; a coincidence in which, if the dark matter candidate has a mass of 100 GeV and couples via the weak force to the SM, then its interaction cross section would be consistent with the observed dark matter abundance. The interaction cross section was also below the direct detection measurements at that time. The simplest WIMP miracle paradigm has now been ruled out by direct detection experiments such as those currently being undertaken at XENON [266] and LUX [267], which means other candidates must be explored.

Many possibilities for the identity and properties of dark matter have been proposed [268–272], but as of yet none have been found. Neutrinos were once considered possible candidates for dark matter [273], but their masses and density are too low, and their relativistic nature early in the universe would have hindered structure formation. Non-particle explanations for dark matter, which constitute alternate gravity theories, have also been proposed, the most well-known being Modified Newtonian Dynamics [274], but these have been mostly unsuccessful in explaining all of the observational phenomena.

The dark matter particles are considered to be stable, to the extent that their lifetime is longer than the age of the universe, because if they decay too quickly they will fail to facilitate successful structure formation. It is also possible that there could be various stable particles present in non-zero densities in the universe, making up components of the dark matter density [275]. A current issue with the cold dark matter model is that simulations predict more satellite galaxies to be present than are observed, which could be a result of small structure formation suppression by dark matter which is warm, rather than cold as usually defined. If the dark matter carries more kinetic energy it can more easily escape gravitational potentials, and hence contribute to the washing out of small scale structures.

One of the most attractive paradigms for describing dark matter is asymmetric dark matter, that is there is a matter-antimatter asymmetry in the dark matter sector too, which may have a similar origin to that in the baryonic sector. This is motivated by the measurement of the ratio of the energy densities of luminous and non-luminous matter,

$$\rho_{\text{DM}} \simeq 5.5\rho_B \ . \tag{1.19}$$

The similarity in these observed densities could indicate a connection between the dynamics and cosmological evolution of visible and dark matter. Many models attempting to explain this ratio have been proposed, and it is an active area of research [237–239, 276–299]. We will consider the possibility of an inflationary cogenesis scenario in Chap. 3.

1.3.4 Neutrino Properties and the Cosmic Neutrino Background

The true nature of neutrinos has been an intriguing mystery since they were first postulated, due to their weakly interacting nature. There have been major advances in our understanding of the properties of neutrinos in the last couple of decades due to the ever improving sensitivity of experiments [300]. The Super-Kamiokande [301–306], Sudbury Neutrino Observatory (SNO) [307–309] and other experiments [310–312], provided the first experimental evidence of neutrino mass and furthered our knowledge of their interaction properties. More recent experiments have begun determining the neutrino mixing angles, and are getting closer to obtaining a measurement of the neutrino $\mathcal{CP}$ violating phase.

One of the main questions regarding neutrinos at present is what are and is the origin of their masses; the mass hierarchy of the neutrino generations is also still unknown. Neutrinos can have either a Dirac mass term $m_\nu \bar{\nu}_L \nu_R$ or a Majorana mass term $m_\nu \nu_L^c \nu_L$, each of which have interesting phenomenological implications. Particularly the Majorana case, because of the lepton violating nature of the mass term, which can not only induce double beta decay, but can have consequences for Baryogenesis. It is possible that there is also other neutrino species, such as sterile neutrinos which do not couple to any of the SM gauge fields, that can have interesting cosmological implications [313].

Another feature of neutrinos which makes them important for early universe cosmology is that, much like photons and the production of the CMB, there is an analogous relic neutrino background; the CνB. This background is produced much earlier than the CMB, and as such could have information encoded within that could unlock many of the current mysteries in particle physics and cosmology. The information could help us gain a greater understanding of the neutrino sector and its role in the early universe. Unfortunately, the low temperature and density of this background today means it is unlikely we will ever be able to directly observe it. There is still hope that it could be indirectly observed, and this is what we shall be exploring in Chap. 5, through the possible effects the CνB could have on gravitational waves.

1.3.5   Gravitational Waves

A determination of the fundamental description of gravity would be revolutionary in physics and have many wide reaching implications. The recent observation of gravitational waves by the LIGO collaboration [314] signals the beginning of the new era of gravitational wave astronomy, allowing the opening of a new window into the fundamental workings of gravity and the universe. To date, the LIGO collaboration has reported the measurement of gravitational waves from several binary black hole merger events [315, 316]. These observations have allowed constraints to be put on potential extensions to general relativity [317–320], and possibly in future, other areas of astrophysics through combined analyses, for example with ANTARES and IceCube [321]. This is a major step towards gaining a greater understanding of the workings of gravity. Future gravitational wave detectors, such as eLISA [322, 323], will also be able to differentiate the polarisations of incoming signals. This will provide information beyond the amplitude and waveform of the waves, and will enable a deeper analysis of the gravitational wave source as well as the fundamental workings of gravity itself. This could be achieved through identification of birefringent propagation effects, which could be smoking guns for certain extensions of General Relativity.

An exciting aspect of such searches is the possible implications for particle physics and early cosmological evolution. Unlike light, for which the early universe plasma was opaque prior to the CMB, it is possible that the gravitational wave remnants from events prior to the formation of the CMB could be observable by future detectors. These observations could provide constraints on early cosmology, and the particle dynamics at that time, that are not achievable with current measurement tools. An example of this shall be discussed in this work with reference to the Cosmic Neutrino Background [5], which could shed light on the true nature of neutrinos and also provide information concerning the origin of the matter-antimatter asymmetry.

1.3.6 Hierarchy Problem in the Standard Model

The hierarchy problem is associated with the unnaturalness of the difference between the apparent fundamental scales of nature, with regards to the SM and gravity, namely the weak and Planck scales [324–331]. This has led to much theoretical anxiety due to the level of fine tuning required to replicate observables if there are no new fundamental scales between the weak and Planck scales. This issue is illustrated by the predicted mass of the Higgs boson, for which the quantum corrections contain quadratic divergences dependent on the scale of new physics. The, recently discovered, Higgs mass has been found to be much lighter than the Planck scale, confirming this disparity in scales.

Supersymmetry has been one of the most studied beyond the SM theories because it can provide a potentially natural solution to the hierarchy problem [332]. The lack of discovery of the Supersymmetry partners has led to the consideration of increasingly complicated Supersymmetry models, which themselves contain issues with fine tuning.

Another solution to the hierarchy problem, which has gained increased interest in recent times, is the introduction of a conformal symmetry into the SM [333]. This model postulates a theory which has no fundamental scales, and that any observed scale is generated dynamically by the spontaneous breaking of the conformal symmetry. The Lagrangian of such a theory contains no dimensionful couplings, and as such has no mass terms. Seeing as the conformal symmetry removes fundamental scales it is a novel way of solving the hierarchy problem. Scale invariant theories have been studied extensively in the past, and have experienced a recent resurgence in interest, and have various interesting implications for the cosmological evolution of the universe [334–364]. In this work we shall consider the application of scale invariance to the inflationary epoch [1].

1.4 Cosmological Implications of Quantum Anomalies

In the following chapters we shall explore the implications of quantum anomalies in the evolutionary history of the universe. Firstly, in Chap. 2, we consider an inflationary mechanism within the setting of a scale invariant theory, providing both a mechanism for inflation and a solution to the hierarchy problem [1]. In Chap. 3, we formulate a generalised version of the inflationary mechanism for Baryogenesis we introduced in [2], which can now produce the correct baryon number density as well as dark matter; through the addition of an anomalous $U(1)_X$ and dark matter fermion to the SM [3]. Chapter 4, considers a new mechanism for Baryogenesis during the reheating epoch, in which we introduce a scalar inflaton and complex scalar baryon which are derivatively coupled [4]. In Chap. 5, we consider a new way to illuminate the properties of the neutrino sector through attempting to constrain the lepton asymmetry carried by the CνB due to effects it can induce with respect to gravitational wave propagation and gravitational instabilities [5].

References

1. N.D. Barrie, A. Kobakhidze, S. Liang, Natural inflation with hidden scale invariance. Phys. Lett. B **756**, 390–393 (2016). https://doi.org/10.1016/j.physletb.2016.03.056
2. N.D. Barrie, A. Kobakhidze, Inflationary baryogenesis in a model with gauged baryon number. JHEP **09**, 163 (2014). https://doi.org/10.1007/JHEP09(2014)163
3. N.D. Barrie, A. Kobakhidze, Generating luminous and dark matter during inflation. Mod. Phys. Lett. A **32**(14), 1750087 (2017). https://doi.org/10.1142/S0217732317500870
4. K. Bamba, N.D. Barrie, A. Sugamoto, T. Takeuchi, K. Yamashita, Ratchet baryogenesis with an analogy to the forced pendulum (2016), arXiv:1610.03268
5. N.D. Barrie, A. Kobakhidze, Gravitational instabilities of the cosmic neutrino background with non-zero lepton number. Phys. Lett. B **772**, 459–463 (2017). https://doi.org/10.1016/j.physletb.2017.07.012
6. N.D. Barrie, A. Kobakhidze, S. Liang, M. Talia, L. Wu, Heavy Leptonium as the origin of the 750 GeV diphoton excess (2016), arXiv:1604.02803
7. N.D. Barrie, A. Kobakhidze, M. Talia, W. Lei, 750 GeV composite axion as the LHC diphoton resonance. Phys. Lett. B **755**, 343–347 (2016). https://doi.org/10.1016/j.physletb.2016.02.010
8. N.D. Barrie, A. Sugamoto, K. Yamashita. Construction of a model of monopolium and its search via multiphoton channels at LHC. *PTEP*, 2016(11), 113B02 (2016). https://doi.org/10.1093/ptep/ptw155
9. C.-N. Yang, R.L. Mills, Conservation of isotopic spin and isotopic gauge invariance. Phys. Rev. **96**, 191–195 (1954). https://doi.org/10.1103/PhysRev.96.191
10. R. Utiyama, Invariant theoretical interpretation of interaction. Phys. Rev. **101**, 1597–1607 (1956). https://doi.org/10.1103/PhysRev.101.1597
11. T.D. Lee, C.-N. Yang, Question of parity conservation in weak interactions. Phys. Rev. **104**, 254–258 (1956). https://doi.org/10.1103/PhysRev.104.254
12. J.S. Schwinger, A theory of the fundamental interactions. Ann. Phys. **2**, 407–434 (1957). https://doi.org/10.1016/0003-4916(57)90015-5
13. C.S. Wu, E. Ambler, R.W. Hayward, D.D. Hoppes, R.P. Hudson, Experimental test of parity conservation in beta decay. Phys. Rev. **105**, 1413–1414 (1957). https://doi.org/10.1103/PhysRev.105.1413
14. S.L. Glashow, Partial symmetries of weak interactions. Nucl. Phys. **22**, 579–588 (1961). https://doi.org/10.1016/0029-5582(61)90469-2
15. J. Goldstone, Field theories with superconductor solutions. Nuovo Cim. **19**, 154–164 (1961). https://doi.org/10.1007/BF02812722
16. Y. Nambu, G. Jona-Lasinio, Dynamical model of elementary particles based on an analogy with superconductivity. 1. Phys. Rev. **122**, 345–358 (1961). https://doi.org/10.1103/PhysRev.122.345

17. Y. Nambu, G. Jona-Lasinio, Dynamical model of elementary particles based on an analogy with superconductivity. 2. Phys. Rev. **124**, 246–254 (1961). https://doi.org/10.1103/PhysRev.124.246

18. J. Goldstone, A. Salam, S. Weinberg, Broken symmetries. Phys. Rev. **127**, 965–970 (1962). https://doi.org/10.1103/PhysRev.127.965

19. Z. Maki, M. Nakagawa, S. Sakata, Remarks on the unified model of elementary particles. Prog. Theor. Phys. **28**, 870–880 (1962). https://doi.org/10.1143/PTP.28.870

20. M. Gell-Mann, Symmetries of baryons and mesons. Phys. Rev. **125**, 1067–1084 (1962). https://doi.org/10.1103/PhysRev.125.1067

21. P.W. Anderson, Plasmons, gauge invariance, and mass. Phys. Rev. **130**, 439–442 (1963). https://doi.org/10.1103/PhysRev.130.439

22. N. Cabibbo, Unitary symmetry and leptonic decays. Phys. Rev. Lett. **10**, 531–533 (1963). https://doi.org/10.1103/PhysRevLett.10.531

23. G. Zweig, An SU(3) model for strong interaction symmetry and its breaking. Version 2, in *Developments in the Quark Theory of Hadrons. 1964–1978*, vol. 1, ed. by D. Lichtenberg, S. Rosen (1964), pp. 22–101

24. M. Gell-Mann, A schematic model of baryons and mesons. Phys. Lett. **8**, 214–215 (1964). https://doi.org/10.1016/S0031-9163(64)92001-3

25. G.S. Guralnik, C.R. Hagen, T.W.B. Kibble, Global conservation laws and massless particles. Phys. Rev. Lett. **13**, 585–587 (1964). https://doi.org/10.1103/PhysRevLett.13.585

26. A. Salam, J. Clive Ward, Electromagnetic and weak interactions. Phys. Lett. **13**, 168–171 (1964). https://doi.org/10.1016/0031-9163(64)90711-5

27. T.W.B. Kibble, Symmetry breaking in nonAbelian gauge theories. Phys. Rev. **155**, 1554–1561 (1967). https://doi.org/10.1103/PhysRev.155.1554

28. S.R. Coleman, J. Mandula, All possible symmetries of the S matrix. Phys. Rev. **159**, 1251–1256 (1967). https://doi.org/10.1103/PhysRev.159.1251

29. S. Weinberg, A model of leptons. Phys. Rev. Lett. **19**, 1264–1266 (1967). https://doi.org/10.1103/PhysRevLett.19.1264

30. A. Salam, Weak and electromagnetic interactions, in *Conference Proceeding*, vol. C680519 (1968), pp. 367–377

31. K.G. Wilson, The renormalization group and strong interactions. Phys. Rev. D **3**, 1818 (1971). https://doi.org/10.1103/PhysRevD.3.1818

32. S.L. Glashow, J. Iliopoulos, L. Maiani, Weak interactions with lepton-hadron symmetry. Phys. Rev. D **2**, 1285–1292 (1970). https://doi.org/10.1103/PhysRevD.2.1285

33. C. Bouchiat, J. Iliopoulos, P. Meyer, An anomaly free version of weinberg's model. Phys. Lett. B **38**, 519–523 (1972). https://doi.org/10.1016/0370-2693(72)90532-1

34. S. Weinberg, Effects of a neutral intermediate boson in semileptonic processes. Phys. Rev. D **5**, 1412–1417 (1972). https://doi.org/10.1103/PhysRevD.5.1412

35. C.G. Bollini, J.J. Giambiagi, Dimensional renormalization: the number of dimensions as a regularizing parameter. Nuovo Cim. B **12**, 20–26 (1972). https://doi.org/10.1007/BF02895558

36. G. Hooft, M.J.G. Veltman, Regularization and renormalization of gauge fields. Nucl. Phys. B **44**, 189–213 (1972). https://doi.org/10.1016/0550-3213(72)90279-9

37. K.G. Wilson, J.B. Kogut, The renormalization group and the epsilon expansion. Phys. Rept. **12**, 75–200 (1974). https://doi.org/10.1016/0370-1573(74)90023-4

38. M. Kobayashi, T. Maskawa, CP violation in the renormalizable theory of weak interaction. Prog. Theor. Phys. **49**, 652–657 (1973). https://doi.org/10.1143/PTP.49.652

39. Jogesh C. Pati, Abdus Salam, Lepton Number as the Fourth Color. Phys. Rev. D **10**, 275–289 (1974). https://doi.org/10.1103/PhysRevD.10.275, https://doi.org/10.1103/PhysRevD.11.703.2. [Erratum: Phys. Rev. D11, 703 (1975)]

40. J. Iliopoulos, The making of the standard theory. Adv. Ser. Direct. High Energy Phys. **26**, 29–59 (2016). https://doi.org/10.1142/9789814733519_0002

41. K.A. Olive et al., Review of particle physics. Chin. Phys. C **38**, 090001 (2014). https://doi.org/10.1088/1674-1137/38/9/090001

42. P.W. Higgs, Broken symmetries and the masses of gauge bosons. Phys. Rev. Lett. **13**, 508–509 (1964). https://doi.org/10.1103/PhysRevLett.13.508

43. P.W. Higgs, Broken symmetries, massless particles and gauge fields. Phys. Lett. **12**, 132–133 (1964). https://doi.org/10.1016/0031-9163(64)91136-9

44. F. Englert, R. Brout, Broken symmetry and the mass of gauge vector mesons. Phys. Rev. Lett. **13**, 321–323 (1964). https://doi.org/10.1103/PhysRevLett.13.321

45. P.W. Higgs, Spontaneous symmetry breakdown without massless bosons. Phys. Rev. **145**, 1156–1163 (1966). https://doi.org/10.1103/PhysRev.145.1156

46. L. Evans, P. Bryant, LHC machine. JINST **3**, S08001 (2008). https://doi.org/10.1088/1748-0221/3/08/S08001

47. G. Aad et al., The ATLAS experiment at the CERN large hadron collider. JINST **3**, S08003 (2008). https://doi.org/10.1088/1748-0221/3/08/S08003

48. G. Aad et al., Observation of a new particle in the search for the standard model higgs boson with the ATLAS detector at the LHC. Phys. Lett. B **716**, 1–29 (2012). https://doi.org/10.1016/j.physletb.2012.08.020

49. S. Chatrchyan et al., The CMS experiment at the CERN LHC. JINST **3**, S08004 (2008). https://doi.org/10.1088/1748-0221/3/08/S08004

50. S. Chatrchyan et al., Observation of a new boson at a mass of 125 GeV with the CMS experiment at the LHC. Phys. Lett. B **716**, 30–61 (2012). https://doi.org/10.1016/j.physletb.2012.08.021

51. S. Weinberg, *The Quantum Theory of Fields. Vol. 1: Foundations* (Cambridge University Press, 2005). ISBN 9780521670531, 9780511252044

52. S. Weinberg, *The Quantum Theory of Fields. Vol. 2: Modern Applications* (Cambridge University Press, 2013). ISBN 9781139632478, 9780521670548, 9780521550024

53. L.H. Ryder. *Quantum Field Theory* (Cambridge University Press, 1996). ISBN 9780521478144, 9781139632393, 9780521237642

54. M.D. Schwartz *Quantum Field Theory and the Standard Model* (Cambridge University Press, 2014). ISBN 1107034736, 9781107034730

55. W.N. Cottingham, D.A. Greenwood, *An Introduction to the Standard Model of Particle Physics* (Cambridge University Press, 2007). ISBN 9780511271366, 9780521852494

56. T. Morii, C.S. Lim, S.N. Mukherjee, *The Physics of the Standard Model and Beyond* (World Scientific, River Edge, USA, 2004)

57. E. Leader, E. Predazzi, An Introduction to gauge theories and modern particle physics. Vol. 2: CP violation, QCD and hard processes. Camb. Monogr. Part. Phys. Nucl. Phys. Cosmol. **4**, 1–464 (1996)

58. E. Leader, E. Predazzi, *An Introduction to Gauge Theories and Modern Particle Physics. Vol. 1: Electroweak Interactions, the New Particles and the Parton Model* (Cambridge University Press, 2011). ISBN 9780511885730, 9780521468404

59. S.R. Coleman, E.J. Weinberg, Radiative corrections as the origin of spontaneous symmetry breaking. Phys. Rev. D **7**, 1888–1910 (1973). https://doi.org/10.1103/PhysRevD.7.1888

60. S. Weinberg, Implications of dynamical symmetry breaking. Phys. Rev. D **13**, 974–996 (1976). https://doi.org/10.1103/PhysRevD.13.974

61. L. Susskind, Dynamics of spontaneous symmetry breaking in the weinberg-salam theory. Phys. Rev. D **20**, 2619–2625 (1979). https://doi.org/10.1103/PhysRevD.20.2619

62. S. Weinberg, Implications of dynamical symmetry breaking: an addendum. Phys. Rev. D **19**, 1277–1280 (1979). https://doi.org/10.1103/PhysRevD.19.1277

63. D.A. Kirzhnits, D. Andrei, Linde. macroscopic consequences of the weinberg model. Phys. Lett. B **42**, 471–474 (1972). https://doi.org/10.1016/0370-2693(72)90109-8

64. L. Dolan, R. Jackiw, Symmetry behavior at finite temperature. Phys. Rev. D **9**, 3320–3341 (1974). https://doi.org/10.1103/PhysRevD.9.3320

65. S. Weinberg, Gauge and global symmetries at high temperature. Phys. Rev. D **9**, 3357–3378 (1974). https://doi.org/10.1103/PhysRevD.9.3357

66. R.D. Peccei, H.R. Quinn, CP conservation in the presence of instantons. Phys. Rev. Lett. **38**, 1440–1443 (1977). https://doi.org/10.1103/PhysRevLett.38.1440

67. M. Lindner, Implications of triviality for the standard model. Z. Phys. C **31**, 295 (1986). https://doi.org/10.1007/BF01479540
68. E. Noether, Invariant variation problems. Transp. Theory Stat. Phys. **1**, 186–207 (1971). https://doi.org/10.1080/00411457108231446
69. S. Rajpoot, Gauge symmetries of electroweak interactions. Int. J. Theor. Phys. **27**, 689 (1988). https://doi.org/10.1007/BF00669312
70. S.L. Adler, Axial vector vertex in spinor electrodynamics. Phys. Rev. **177**, 2426–2438 (1969). https://doi.org/10.1103/PhysRev.177.2426
71. J.S. Bell, R. Jackiw, A PCAC puzzle: pi0 gt; gamma gamma in the sigma model. Nuovo Cim. A **60**, 47–61 (1969). https://doi.org/10.1007/BF02823296
72. G. Hooft, Symmetry breaking through Bell-Jackiw anomalies. Phys. Rev. Lett. **37**, 8–11 (1976). https://doi.org/10.1103/PhysRevLett.37.8
73. J. Preskill, Gauge anomalies in an effective field theory. Ann. Phys. **210**, 323–379 (1991). https://doi.org/10.1016/0003-4916(91)90046-B
74. M.B. Green, J.H. Schwarz, Anomaly cancellation in supersymmetric D = 10 gauge theory and superstring theory. Phys. Lett. B **149**, 117–122 (1984). https://doi.org/10.1016/0370-2693(84)91565-X
75. P.F. Perez, T. Han, T. Li, M.J. Ramsey-Musolf, Leptoquarks and neutrino masses at the LHC. Nucl. Phys. B **819**, 139–176 (2009). https://doi.org/10.1016/j.nuclphysb.2009.04.009
76. P.V. Dong, H.N. Long, A simple model of gauged lepton and baryon charges. Phys. Int. **6**(1), 23–32 (2010). https://doi.org/10.3844/pisp.2015.23.32
77. M. Duerr, P.F. Perez, M.B. Wise, Gauge theory for baryon and lepton numbers with Leptoquarks. Phys. Rev. Lett. **110**, 231801 (2013). https://doi.org/10.1103/PhysRevLett.110.231801
78. E.C.G. Stueckelberg, Interaction energy in electrodynamics and in the field theory of nuclear forces. Helv. Phys. Acta **11**, 225–244 (1938). https://doi.org/10.5169/seals-110852
79. S. Weinberg, *Gravitation and Cosmology* (Wiley, New York, 1972). ISBN 0471925675, 9780471925675
80. S. Dodelson, *Modern Cosmology* (Academic Press, Amsterdam, 2003). ISBN 9780122191411
81. M. Trodden, S.M. Carroll, TASI lectures: introduction to cosmology, in *Progress in String Theory. Proceedings, Summer School, TASI 2003, Boulder, USA, June 2–27, 2003* (2004), pp. 703–793
82. V. Mukhanov, *Physical Foundations of Cosmology* (Cambridge University Press, Oxford, 2005). ISBN 0521563984, 9780521563987
83. A.A. Penzias, R.W. Wilson, A measurement of excess antenna temperature at 4080-Mc/s. Astrophys. J. **142**, 419–421 (1965). https://doi.org/10.1086/148307
84. J.C. Mather et al., A preliminary measurement of the cosmic microwave background spectrum by the Cosmic Background Explorer (COBE) satellite. Astrophys. J. **354**, L37–L40 (1990). https://doi.org/10.1086/185717
85. G.F. Smoot et al., Structure in the COBE differential microwave radiometer first year maps. Astrophys. J. **396**, L1–L5 (1992). https://doi.org/10.1086/186504
86. D.N. Spergel et al., First year Wilkinson Microwave Anisotropy Probe (WMAP) observations: determination of cosmological parameters. Astrophys. J. Suppl. **148**, 175–194 (2003). https://doi.org/10.1086/377226
87. D.N. Spergel et al., Wilkinson Microwave Anisotropy Probe (WMAP) three year results: implications for cosmology. Astrophys. J. Suppl. **170**, 377 (2007). https://doi.org/10.1086/513700
88. E. Komatsu et al., Five-year Wilkinson Microwave Anisotropy Probe (WMAP) observations: cosmological interpretation. Astrophys. J. Suppl. **180**, 330–376 (2009). https://doi.org/10.1088/0067-0049/180/2/330
89. E. Komatsu et al., Seven-year Wilkinson Microwave Anisotropy Probe (WMAP) observations: cosmological interpretation. Astrophys. J. Suppl. **192**, 18 (2011). https://doi.org/10.1088/0067-0049/192/2/18

90. C.L. Bennett et al., Nine-year Wilkinson Microwave Anisotropy Probe (WMAP) observations: final maps and results. Astrophys. J. Suppl. **208**, 20 (2013). https://doi.org/10.1088/0067-0049/208/2/20
91. G. Hinshaw et al., Nine-year Wilkinson Microwave Anisotropy Probe (WMAP) observations: cosmological parameter results. Astrophys. J. Suppl. **208**, 19 (2013). https://doi.org/10.1088/0067-0049/208/2/19
92. P.A.R. Ade et al., Planck 2013 results. I. Overview of products and scientific results. Astron. Astrophys. **571**, A1 (2014). https://doi.org/10.1051/0004-6361/201321529
93. R. Adam et al., Planck 2015 results. I. Overview of products and scientific results. Astron. Astrophys. **594**, A1 (2016). https://doi.org/10.1051/0004-6361/201527101
94. P.A.R. Ade et al., Planck 2013 results. I. Overview of products and scientific results. Astron. Astrophys. **571**, A1 (2014a). https://doi.org/10.1051/0004-6361/201321529
95. J.B. Hartle, *An Introduction to Einstein's General Relativity* (Addison-Wesley, San Francisco, USA, 2003)
96. E. Hubble, A relation between distance and radial velocity among extra-galactic nebulae. Proc. Nat. Acad. Sci. **15**, 168–173 (1929). https://doi.org/10.1073/pnas.15.3.168
97. G. Gamow, Expanding universe and the origin of elements. Phys. Rev. **70**, 572–573 (1946). https://doi.org/10.1103/PhysRev7.0.572
98. R.A. Alpher, H. Bethe, G. Gamow, The origin of chemical elements. Phys. Rev. **73**, 803–804 (1948). https://doi.org/10.1103/PhysRev.73.803
99. R.V. Wagoner, W.A. Fowler, F. Hoyle, On the synthesis of elements at very high temperatures. Astrophys. J. **148**, 3–49 (1967). https://doi.org/10.1086/149126
100. D.N. Schramm, R.V. Wagoner, Element production in the early universe. Ann. Rev. Nucl. Part. Sci. **27**, 37–74 (1977). https://doi.org/10.1146/annurev.ns.27.120177.000345
101. J.-M. Yang, M.S. Turner, G. Steigman, D.N. Schramm, K.A. Olive, Primordial nucleosynthesis: a critical comparison of theory and observation. Astrophys. J. **281**, 493–511 (1984). https://doi.org/10.1086/162123
102. A. Merchant Boesgaard , G. Steigman, Big bang nucleosynthesis: theories and observations. Ann. Rev. Astron. Astrophys. **23**, 319–378 (1985). https://doi.org/10.1146/annurev.aa.23.090185.001535
103. C.J. Copi, D.N. Schramm, M.S. Turner, Big bang nucleosynthesis and the baryon density of the universe. Science **267**, 192–199 (1995). https://doi.org/10.1126/science.7809624
104. G. Steigman, Primordial nucleosynthesis in the precision cosmology era. Ann. Rev. Nucl. Part. Sci. **57**, 463–491 (2007). https://doi.org/10.1146/annurev.nucl.56.080805.140437
105. F. Iocco, G. Mangano, G. Miele, O. Pisanti, P.D. Serpico, Primordial nucleosynthesis: from precision cosmology to fundamental physics. Phys. Rept. **472**, 1–76 (2009). https://doi.org/10.1016/j.physrep.2009.02.002
106. R.H. Cyburt, B.D. Fields, K.A. Olive, An update on the big bang nucleosynthesis prediction for Li-7: the problem worsens. JCAP **0811**, 012 (2008). https://doi.org/10.1088/1475-7516/2008/11/012
107. R.H. Cyburt, B.D. Fields, K.A. Olive, T.-H. Yeh, Big bang nucleosynthesis: 2015. Rev. Mod. Phys. **88**, 015004 (2016). https://doi.org/10.1103/RevModPhys.88.015004
108. D. Baumann, On the quantum origin of structure in the inflationary universe (2007). http://inspirehep.net/record/827549
109. A.H. Guth, The inflationary universe: a possible solution to the horizon and flatness problems. Phys. Rev. D **23**, 347–356 (1981). https://doi.org/10.1103/PhysRevD.23.347
110. A.A. Starobinsky, A new type of isotropic cosmological models without singularity. Phys. Lett. B **91**, 99–102 (1980). https://doi.org/10.1016/0370-2693(80)90670-X
111. D. Kazanas, Dynamics of the universe and spontaneous symmetry breaking. Astrophys. J. **241**, L59–L63 (1980). https://doi.org/10.1086/183361
112. K. Sato, First order phase transition of a vacuum and expansion of the universe. Mon. Not. Roy. Astron. Soc. **195**, 467–479 (1981)
113. A.D. Linde, A new inflationary universe scenario: a possible solution of the horizon, flatness, homogeneity, isotropy and primordial monopole problems. Phys. Lett. B **108**, 389–393 (1982). https://doi.org/10.1016/0370-2693(82)91219-9

114. A.D. Linde, Coleman-weinberg theory and a new inflationary universe scenario. Phys. Lett. B **114**, 431–435 (1982). https://doi.org/10.1016/0370-2693(82)90086-7

115. A. Albrecht, P.J. Steinhardt, Cosmology for grand unified theories with radiatively induced symmetry breaking. Phys. Rev. Lett. **48**, 1220–1223 (1982). https://doi.org/10.1103/PhysRevLett.48.1220

116. A.H. Guth, S.Y. Pi, Fluctuations in the new inflationary universe. Phys. Rev. Lett. **49**, 1110–1113 (1982). https://doi.org/10.1103/PhysRevLett.49.1110

117. M.S. Turner, Coherent scalar field oscillations in an expanding universe. Phys. Rev. D **28**, 1243 (1983). https://doi.org/10.1103/PhysRevD.28.1243

118. J.M. Bardeen, P.J. Steinhardt, M.S. Turner, Spontaneous creation of almost scale-free density perturbations in an inflationary universe. Phys. Rev. D **28**, 679 (1983). https://doi.org/10.1103/PhysRevD.28.679

119. A.D. Linde, The inflationary universe. Rept. Prog. Phys. **47**, 925–986 (1984). https://doi.org/10.1088/0034-4885/47/8/002

120. R.H. Brandenberger, Quantum field theory methods and inflationary universe models. Rev. Mod. Phys. **57**, 1 (1985). https://doi.org/10.1103/RevModPhys.57.1

121. A.D. Linde, Eternal chaotic inflation. Mod. Phys. Lett. A **1**, 81 (1986). https://doi.org/10.1142/S0217732386000129

122. D. Baumann, Inflation, in *Physics of the Large and the Small, TASI 2009, Proceedings of the Theoretical Advanced Study Institute in Elementary Particle Physics, Boulder, Colorado, USA, 1–26 June 2009* (2011), pp. 523–686. https://doi.org/10.1142/9789814327183_0010

123. G. Hooft, Magnetic monopoles in unified gauge theories. Nucl. Phys. B **79**, 276–284 (1974). https://doi.org/10.1016/0550-3213(74)90486-6

124. Y.B. Zeldovich, M.Y. Khlopov, On the concentration of relic magnetic monopoles in the universe. Phys. Lett. B **79**, 239–241 (1978). https://doi.org/10.1016/0370-2693(78)90232-0

125. J. Preskill, Cosmological production of superheavy magnetic monopoles. Phys. Rev. Lett. **43**, 1365 (1979). https://doi.org/10.1103/PhysRevLett.43.1365

126. V.F. Mukhanov, G.V. Chibisov, Quantum fluctuations and a nonsingular universe. JETP Lett. **33**, 532–535 (1981). [Pisma Zh. Eksp. Teor. Fiz.33,549(1981)]

127. K.A. Olive, Inflation. Phys. Rept. **190**, 307–403 (1990). https://doi.org/10.1016/0370-1573(90)90144-Q

128. E.W. Kolb, M.S. Turner, The early universe. Front. Phys. **69**, 1–547 (1990)

129. A. Riotto, Inflation and the theory of cosmological perturbations, in *Astroparticle Physics and Cosmology. Proceedings: Summer School, Trieste, Italy, Jun 17–Jul 5 2002* (2002), pp. 317–413

130. R.H. Brandenberger, Lectures on the theory of cosmological perturbations. Lect. Notes Phys. **646**, 127–167 (2004)

131. A.D. Linde, Particle physics and inflationary cosmology. Contemp. Concepts Phys. **5**, 1–362 (1990)

132. K.A. Malik, D. Wands, Cosmological perturbations. Phys. Rept. **475**, 1–51 (2009). https://doi.org/10.1016/j.physrep.2009.03.001

133. R.H. Brandenberger, Alternatives to the inflationary paradigm of structure formation. Int. J. Mod. Phys. Conf. Ser. **01**, 67–79 (2011). https://doi.org/10.1142/S2010194511000109

134. R.H. Brandenberger, Cosmology of the very early universe. AIP Conf. Proc. **1268**, 3–70 (2010). https://doi.org/10.1063/1.3483879

135. R.H. Brandenberger, Unconventional cosmology. Lect. Notes Phys. **863**, 333 (2013). https://doi.org/10.1007/978-3-642-33036-0_12

136. F. Lucchin, S. Matarrese, Power law inflation. Phys. Rev. D **32**, 1316 (1985). https://doi.org/10.1103/PhysRevD.32.1316

137. K. Freese, J.A. Frieman, A.V. Olinto, Natural inflation with pseudo-Nambu-Goldstone bosons. Phys. Rev. Lett. **65**, 3233–3236 (1990). https://doi.org/10.1103/PhysRevLett.65.3233

138. J.D. Barrow, K. Maeda, Extended inflationary universes. Nucl. Phys. B **341**, 294–308 (1990). https://doi.org/10.1016/0550-3213(90)90272-F

139. A.L. Berkin, K.-I. Maeda, Inflation in generalized Einstein theories. Phys. Rev. D **44**, 1691–1704 (1991). https://doi.org/10.1103/PhysRevD.44.1691

140. A.D. Linde, Axions in inflationary cosmology. Phys. Lett. B **259**, 38–47 (1991). https://doi.org/10.1016/0370-2693(91)90130-I

141. L.F. Abbott, E. Farhi, M.B. Wise, Particle production in the new inflationary cosmology. Phys. Lett. **B117**, 29 (1982). https://doi.org/10.1016/0370-2693(82)90867-X

142. S.W. Hawking, The development of irregularities in a single bubble inflationary universe. Phys. Lett. B **115**, 295 (1982). https://doi.org/10.1016/0370-2693(82)90373-2

143. A.A. Starobinsky, Dynamics of phase transition in the new inflationary universe scenario and generation of perturbations. Phys. Lett. B **117**, 175–178 (1982). https://doi.org/10.1016/0370-2693(82)90541-X

144. F.C. Adams, J. Richard Bond, K. Freese, J.A. Frieman, A.V. Olinto, Natural inflation: particle physics models, power law spectra for large scale structure, and constraints from COBE. Phys. Rev. D **47**, 426–455 (1993). https://doi.org/10.1103/PhysRevD.47.426

145. D.A. Linde, Chaotic inflation. Phys. Lett. B **129**, 177–181 (1983). https://doi.org/10.1016/0370-2693(83)90837-7

146. A.D. Linde, Hybrid inflation. Phys. Rev. D **49**, 748–754 (1994). https://doi.org/10.1103/PhysRevD.49.748

147. E.J. Copeland, A.R. Liddle, D.H. Lyth, E.D. Stewart, D. Wands, False vacuum inflation with Einstein gravity. Phys. Rev. D **49**, 6410–6433 (1994). https://doi.org/10.1103/PhysRevD.49.6410

148. A. Berera, Warm inflation. Phys. Rev. Lett. **75**, 3218–3221 (1995). https://doi.org/10.1103/PhysRevLett.75.3218

149. P. Binetruy, G.R. Dvali, D term inflation. Phys. Lett. B **388**, 241–246 (1996). https://doi.org/10.1016/S0370-2693(96)01083-0

150. G.R. Dvali, S.H. Henry Tye, Brane inflation. Phys. Lett. B **450**, 72–82 (1999). https://doi.org/10.1016/S0370-2693(99)00132-X

151. D.H. Lyth, A. Riotto, Particle physics models of inflation and the cosmological density perturbation. Phys. Rept. **314**, 1–146 (1999). https://doi.org/10.1016/S0370-1573(98)00128-8

152. A.R. Liddle, A. Mazumdar, F.E. Schunck, Assisted inflation. Phys. Rev. D **58**, 061301 (1998). https://doi.org/10.1103/PhysRevD.58.061301

153. C. Armendariz-Picon, T. Damour, V.F. Mukhanov. k-inflation. Phys. Lett. B **458**, 209–218 (1999). https://doi.org/10.1016/S0370-2693(99)00603-6

154. A. Mazumdar, Extra dimensions and inflation. Phys. Lett. B **469**, 55–60 (1999). https://doi.org/10.1016/S0370-2693(99)01256-3

155. L. Boubekeur, D.H. Lyth, Hilltop inflation. JCAP **0507**, 010 (2005). https://doi.org/10.1088/1475-7516/2005/07/010

156. S. Dimopoulos, S. Kachru, J. McGreevy, J.G. Wacker, N-flation. JCAP **0808**, 003 (2008). https://doi.org/10.1088/1475-7516/2008/08/003

157. J. Martin, C. Ringeval, V. Vennin, Encyclopædia Inflationaris. Phys. Dark Univ. **5–6**, 75–235 (2014). https://doi.org/10.1016/j.dark.2014.01.003

158. L.F. Abbott, M.B. Wise, Constraints on generalized inflationary cosmologies. Nucl. Phys. B **244**, 541–548 (1984). https://doi.org/10.1016/0550-3213(84)90329-8

159. F.C. Adams, K. Freese, A.H. Guth, Constraints on the scalar field potential in inflationary models. Phys. Rev. D **43**, 965–976 (1991). https://doi.org/10.1103/PhysRevD.43.965

160. D.H. Lyth, What would we learn by detecting a gravitational wave signal in the cosmic microwave background anisotropy? Phys. Rev. Lett. **78**, 1861–1863 (1997). https://doi.org/10.1103/PhysRevLett.78.1861

161. D.N. Spergel, M. Zaldarriaga, CMB polarization as a direct test of inflation. Phys. Rev. Lett. **79**, 2180–2183 (1997). https://doi.org/10.1103/PhysRevLett.79.2180

162. A.R. Liddle, S.M. Leach, How long before the end of inflation were observable perturbations produced? Phys. Rev. D **68**, 103503 (2003). https://doi.org/10.1103/PhysRevD.68.103503

163. L. Alabidi, D.H. Lyth, Inflation models and observation. JCAP **0605**, 016 (2006). https://doi.org/10.1088/1475-7516/2006/05/016

164. J.L. Cook, L. Sorbo, Particle production during inflation and gravitational waves detectable by ground-based interferometers. Phys. Rev. D **85**, 023534 (2012). https://doi.org/10.1103/PhysRevD.86.069901, https://doi.org/10.1103/PhysRevD.85.023534. [Erratum: Phys. Rev. D86, 069901 (2012)]

165. D.H. Lyth, The CMB modulation from inflation. JCAP **1308**, 007 (2013). https://doi.org/10.1088/1475-7516/2013/08/007

166. J. Martin, C. Ringeval, R. Trotta, V. Vennin, The best inflationary models after planck. JCAP **1403**, 039 (2014). https://doi.org/10.1088/1475-7516/2014/03/039

167. D. Roest, Universality classes of inflation. JCAP **1401**, 007 (2014). https://doi.org/10.1088/1475-7516/2014/01/007

168. M. Galante, R. Kallosh, A. Linde, D. Roest, Unity of cosmological inflation attractors. Phys. Rev. Lett. **114**(14), 141302 (2015). https://doi.org/10.1103/PhysRevLett.114.141302

169. P. Binetruy, E. Kiritsis, J. Mabillard, M. Pieroni, C. Rosset, Universality classes for models of inflation. JCAP **1504**(04), 033 (2015). https://doi.org/10.1088/1475-7516/2015/04/033

170. V. Domcke, M. Pieroni, P. Binétruy, Primordial gravitational waves for universality classes of pseudoscalar inflation. JCAP **1606**, 031 (2016). https://doi.org/10.1088/1475-7516/2016/06/031

171. H.-Y. Chiu, Symmetry between particle and anti-particle populations in the universe. Phys. Rev. Lett. **17**, 712 (1966). https://doi.org/10.1103/PhysRevLett.17.712

172. G. Steigman, Observational tests of antimatter cosmologies. Ann. Rev. Astron. Astrophys. **14**, 339–372 (1976). https://doi.org/10.1146/annurev.aa.14.090176.002011

173. A.Y. Ignatiev, N.V. Krasnikov, V.A. Kuzmin, A.N. Tavkhelidze, Universal CP noninvariant superweak interaction and baryon asymmetry of the universe. Phys. Lett. B **76**, 436–438 (1978). https://doi.org/10.1016/0370-2693(78)90900-0

174. D. Toussaint, S.B. Treiman, F. Wilczek, A. Zee, Matter-antimatter accounting, thermodynamics, and black hole radiation. Phys. Rev. D **19**, 1036–1045 (1979). https://doi.org/10.1103/PhysRevD.19.1036

175. S. Dimopoulos, L. Susskind, On the baryon number of the universe. Phys. Rev. D **18**, 4500–4509 (1978). https://doi.org/10.1103/PhysRevD.18.4500

176. M. Yoshimura, Origin of cosmological baryon asymmetry. Phys. Lett. B **88**, 294–298 (1979). https://doi.org/10.1016/0370-2693(79)90471-4

177. S. Weinberg, Cosmological production of baryons. Phys. Rev. Lett. **42**, 850–853 (1979). https://doi.org/10.1103/PhysRevLett.42.850

178. E.W. Kolb, S. Wolfram, Baryon number generation in the early universe. Nucl. Phys. B **172**, 224 (1980). https://doi.org/10.1016/0550-3213(80)90167-4, https://doi.org/10.1016/0550-3213(82)90012-8. [Erratum: Nucl. Phys. B195, 542 (1982)]

179. A.D. Dolgov, Y.B. Zeldovich, Cosmology and elementary particles. Rev. Mod. Phys. **53**, 1–41 (1981). https://doi.org/10.1103/RevModPhys.53.1

180. A.D. Dolgov, NonGUT baryogenesis. Phys. Rept. **222**, 309–386 (1992). https://doi.org/10.1016/0370-1573(92)90107-B

181. W. Buchmuller, T. Yanagida, Baryogenesis and the scale of B-L breaking. Phys. Lett. B **302**, 240–244 (1993). https://doi.org/10.1016/0370-2693(93)90391-T

182. W. Buchmuller, T. Yanagida, Quark lepton mass hierarchies and the baryon asymmetry. Phys. Lett. B **445**, 399–402 (1999). https://doi.org/10.1016/S0370-2693(98)01480-4

183. W. Buchmuller, Some aspects of baryogenesis and lepton number violation, in *Recent Developments in Particle Physics and Cosmology: Proceedings. NATO ASI 2000. Cascais, Portugal, July 26–Jul 7, 2000* (2000), pp. 281–314

184. W. Buchmuller, P. Di Bari, M. Plumacher, Cosmic microwave background, matter-antimatter asymmetry and neutrino masses. Nucl. Phys. B **643**, 367–390 (2002). https://doi.org/10.1016/S0550-3213(02)00737-X, https://doi.org/10.1016/j.nuclphysb.2007.11.030. [Erratum: Nucl. Phys. B793, 362 (2008)]

185. P.A.M. Dirac, A theory of electrons and protons. Proc. Roy. Soc. Lond. A **126**, 360 (1930). https://doi.org/10.1098/rspa.1930.0013

186. J.H. Christenson, J.W. Cronin, V.L. Fitch, R. Turlay, Evidence for the 2 pi Decay of the k(2)0 Meson. Phys. Rev. Lett. **13**, 138–140 (1964). https://doi.org/10.1103/PhysRevLett.13.138

187. G. D'Ambrosio, G. Isidori, CP violation in kaon decays. Int. J. Mod. Phys. A **13**, 1–94 (1998). https://doi.org/10.1142/S0217751X98000020

188. R. Aaij et al., Evidence for CP violation in time-integrated $D^0 \to h^- h^+$ decay rates. Phys. Rev. Lett. **108**, 111602 (2012). https://doi.org/10.1103/PhysRevLett.108.129903, https://doi.org/10.1103/PhysRevLett.108.111602

189. R. Aaij et al., First observation of CP violation in the decays of B_s^0 mesons. Phys. Rev. Lett. **110**, 221601 (2013). https://doi.org/10.1103/PhysRevLett.110.221601

190. G. Borissov, R. Fleischer, M.-H. Schune, Rare decays and CP violation in the bs system. Annu. Rev. Nucl. Part. Sci. **63**(1), null (2013). https://doi.org/10.1146/annurev-nucl-102912-144527

191. S.P. Ahlen, S. Barwick, J.J. Beatty, C.R. Bower, G. Gerbier et al., New limit on the low-energy anti-proton / proton ratio in the galactic cosmic radiation. Phys. Rev. Lett. **61**, 145–148 (1988). https://doi.org/10.1103/PhysRevLett.61.145

192. J. Alcaraz et al., Search for anti-helium in cosmic rays. Phys. Lett. B **461**, 387–396 (1999). https://doi.org/10.1016/S0370-2693(99)00874-6

193. P.A.R. Ade et al., Planck 2015 results. XIII. Cosmological parameters. Astron. Astrophys. **594**, A13 (2016). https://doi.org/10.1051/0004-6361/201525830

194. J. Beringer et al., Review of particle physics (RPP). Phys. Rev. D **86**, 010001 (2012). https://doi.org/10.1103/PhysRevD.86.010001

195. V. Simha, G. Steigman, Constraining the early-universe baryon density and expansion rate. JCAP **0806**, 016 (2008). https://doi.org/10.1088/1475-7516/2008/06/016

196. G. Steigman, Primordial nucleosynthesis: the predicted and observed abundances and their consequences. PoS **NICXI**, 001 (2010)

197. B.D. Fields, P. Molaro, S. Sarkar, Big-bang nucleosynthesis. Chin. Phys. C **38**, 339–344 (2014)

198. F.W. Stecker, D.L. Morgan, J. Bredekamp, Possible evidence for the existence of antimatter on a cosmological scale in the universe. Phys. Rev. Lett. **27**, 1469–1472 (1971). https://doi.org/10.1103/PhysRevLett.27.1469

199. S. Dodelson, L.M. Widrow, Baryogenesis in a baryon symmetric universe. Phys. Rev. D **42**, 326–342 (1990). https://doi.org/10.1103/PhysRevD.42.326

200. S. Dodelson, L.M. Widrow, Baryon symmetric baryogenesis. Phys. Rev. Lett. **64**, 340–343 (1990). https://doi.org/10.1103/PhysRevLett.64.340

201. D.L. Morgan, V.W. Hughes, Atomic processes involved in matter-antimatter annihilation. Phys. Rev. D **2**, 1389–1399 (1970). https://doi.org/10.1103/PhysRevD.2.1389

202. L. Canetti, M. Drewes, M. Shaposhnikov, Matter and antimatter in the universe. New J. Phys. **14**, 095012 (2012). https://doi.org/10.1088/1367-2630/14/9/095012

203. A.G. Cohen, A. De, Rujula, S.L. Glashow, A matter-antimatter universe? Astrophys. J. **495**, 539–549 (1998). https://doi.org/10.1086/305328

204. A.G. Cohen, A. De Rujula, Scars on the CBR? (1997), arXiv:astro-ph/9709132

205. A.D. Sakharov, Violation of CP Invariance, c Asymmetry, and Baryon Asymmetry of the Universe. Pisma Zh. Eksp. Teor. Fiz. **5**, 32–35 (1967). https://doi.org/10.1070/PU1991v034n05ABEH002497. [Usp. Fiz. Nauk 161, 61 (1991)]

206. A.D. Linde, Phase transitions in gauge theories and cosmology. Rept. Prog. Phys. **42**, 389 (1979). https://doi.org/10.1088/0034-4885/42/3/001

207. V.A. Kuzmin, V.A. Rubakov, M.E. Shaposhnikov, On the anomalous electroweak baryon number nonconservation in the early universe. Phys. Lett. B **155**, 36 (1985). https://doi.org/10.1016/0370-2693(85)91028-7

208. M.E. Shaposhnikov, Structure of the high temperature gauge ground state and electroweak production of the baryon asymmetry. Nucl. Phys. B **299**, 797–817 (1988). https://doi.org/10.1016/0550-3213(88)90373-2

209. M.E. Shaposhnikov, Baryon asymmetry of the universe in standard electroweak theory. Nucl. Phys. B **287**, 757–775 (1987). https://doi.org/10.1016/0550-3213(87)90127-1

210. N. Turok, J. Zadrozny, Dynamical generation of baryons at the electroweak transition. Phys. Rev. Lett. **65**, 2331–2334 (1990). https://doi.org/10.1103/PhysRevLett.65.2331

211. Michael Dine, Patrick Huet, Robert L. Singleton, Jr., Baryogenesis at the electroweak scale. Nucl. Phys. B **375**, 625–648 (1992). https://doi.org/10.1016/0550-3213(92)90113-P

212. G.W. Anderson, L.J. Hall, The electroweak phase transition and baryogenesis. Phys. Rev. D **45**, 2685–2698 (1992). https://doi.org/10.1103/PhysRevD.45.2685

213. G.W. Anderson, Remarks on the electroweak phase transition, in *1st Yale-Texas Workshop on Baryon Number Violation at the Electroweak Scale New Haven, Connecticut, March 19–21, 1992* (1992), pp. 0134–143

214. A.G. Cohen, D.B. Kaplan, A.E. Nelson, Progress in electroweak baryogenesis. Ann. Rev. Nucl. Part. Sci. **43**, 27–70 (1993). https://doi.org/10.1146/annurev.ns.43.120193.000331

215. G.R. Farrar, M.E. Shaposhnikov, Baryon asymmetry of the universe in the standard electroweak theory. Phys. Rev. D **50**, 774 (1994). https://doi.org/10.1103/PhysRevD.50.774

216. J.M. Cline, Recent progress in electroweak baryogenesis, in *Strong and Electroweak Matter 1998. Proceedings, Conference, SEWM 1998, Copenhagen, Denmark, December 2–5, 1998* (1998), pp. 70–80

217. M. Trodden, Electroweak baryogenesis. Rev. Mod. Phys. **71**, 1463–1500 (1999). https://doi.org/10.1103/RevModPhys.71.1463

218. R.N. Mohapatra, G. Senjanovic, Broken symmetries at high temperature. Phys. Rev. D **20**, 3390–3398 (1979). https://doi.org/10.1103/PhysRevD.20.3390

219. K. Sato, Cosmological baryon number domain structure and the first order phase transition of a vacuum. Phys. Lett. B **99**, 66–70 (1981). https://doi.org/10.1016/0370-2693(81)90805-4

220. N. Blinov, Phase transitions: applications to physics beyond the standard model. Ph.D. thesis, British Columbia University, 2015

221. M.B. Gavela, P. Hernandez, J. Orloff, O. Pene, Standard model CP violation and baryon asymmetry. Mod. Phys. Lett. A **9**, 795–810 (1994). https://doi.org/10.1142/S0217732394000629

222. M. Dine, R. Leigh, P. Huet, A. Linde, D. Linde, Towards the theory of the electroweak phase transition. Phys. Rev. D **46**, 550–571 (1992). https://doi.org/10.1103/PhysRevD.46.550

223. K. Kajantie, M. Laine, K. Rummukainen, M.E. Shaposhnikov, Is there a hot electroweak phase transition at m(H) larger or equal to m(W)? Phys. Rev. Lett. **77**, 2887–2890 (1996). https://doi.org/10.1103/PhysRevLett.77.2887

224. M. Dine, A. Kusenko, The origin of the matter-antimatter asymmetry. Rev. Mod. Phys. **76**, 1 (2003). https://doi.org/10.1103/RevModPhys.76.1

225. A. Riotto, Theories of baryogenesis, in *Proceedings, Summer School in High-energy Physics and Cosmology: Trieste, Italy, June 29–July 17, 1998* (1998), pp. 326–436

226. A. Riotto, M. Trodden, Recent progress in baryogenesis. Ann. Rev. Nucl. Part. Sci. **49**, 35–75 (1999). https://doi.org/10.1146/annurev.nucl.49.1.35

227. W. Buchmuller, S. Fredenhagen, Elements of baryogenesis, in *Current Topics in Astrofundamental Physics: The Cosmic Microwave Background. Proceedings, NATO Advanced Study Institute, 8th Course, Erice, Italy, December 5–16, 1999* (2000), pp. 17–35

228. J.M. Cline, Baryogenesis, in *Les Houches Summer School-Session 86: Particle Physics and Cosmology: The Fabric of Spacetime Les Houches, France, July 31–August 25, 2006* (2006)

229. S.N. Gninenko, D.S. Gorbunov, M.E. Shaposhnikov, Search for GeV-scale sterile neutrinos responsible for active neutrino oscillations and baryon asymmetry of the Universe. Adv. High Energy Phys. **2012**, 718259 (2012). https://doi.org/10.1155/2012/718259

230. W.-M. Yang, A model of four generation fermions and cold dark matter and matter-antimatter asymmetry. Phys. Rev. D **87**, 095003 (2013). https://doi.org/10.1103/PhysRevD.87.095003

231. I. Affleck, M. Dine, A new mechanism for baryogenesis. Nucl. Phys. B **249**, 361 (1985). https://doi.org/10.1016/0550-3213(85)90021-5

232. R. Allahverdi, A. Mazumdar, A mini review on Affleck-Dine baryogenesis. New J. Phys. **14**, 125013 (2012). https://doi.org/10.1088/1367-2630/14/12/125013

233. K. Benakli, S. Davidson, Baryogenesis in models with a low quantum gravity scale. Phys. Rev. D **60**, 025004 (1999). https://doi.org/10.1103/PhysRevD.60.025004

234. M. Yoshimura, Unified gauge theories and the baryon number of the universe. Phys. Rev. Lett. **41**, 281–284 (1978). https://doi.org/10.1103/PhysRevLett.41.281. [Erratum: Phys. Rev. Lett. 42, 746 (1979)]
235. E.W. Kolb, M.S. Turner, Grand unified theories and the origin of the baryon asymmetry. Ann. Rev. Nucl. Part. Sci. **33**, 645–696 (1983). https://doi.org/10.1146/annurev.ns.33.120183. 003241
236. R. Allahverdi, B. Dutta, K. Sinha, Cladogenesis: baryon-dark matter coincidence from branchings in moduli decay. Phys. Rev. D **83**, 083502 (2011). https://doi.org/10.1103/PhysRevD. 83.083502
237. M.R. Buckley, L. Randall, Xogenesis. JHEP **09**, 009 (2011). https://doi.org/10.1007/ JHEP09(2011)009
238. J. Shelton, K.M. Zurek, Darkogenesis: a baryon asymmetry from the dark matter sector. Phys. Rev. D **82**, 123512 (2010). https://doi.org/10.1103/PhysRevD.82.123512
239. M. Blennow, B. Dasgupta, E. Fernandez-Martinez, N. Rius, Aidnogenesis via leptogenesis and dark sphalerons. JHEP **03**, 014 (2011). https://doi.org/10.1007/JHEP03(2011)014
240. L. Canetti, M. Drewes, T. Frossard, M. Shaposhnikov, Dark matter, baryogenesis and neutrino oscillations from right handed neutrinos. Phys. Rev. D **87**, 093006 (2013). https://doi.org/10. 1103/PhysRevD.87.093006
241. S.M. Boucenna, S. Morisi, Theories relating baryon asymmetry and dark matter: a mini review. Front. Phys. **1**, 33 (2014). https://doi.org/10.3389/fphy.2013.00033
242. C. Cheung, Y. Zhang, Electroweak cogenesis. JHEP **1309**, 002 (2013). https://doi.org/10. 1007/JHEP09(2013)002
243. M.A. Luty, Baryogenesis via leptogenesis. Phys. Rev. D **45**, 455–465 (1992). https://doi.org/ 10.1103/PhysRevD.45.455
244. A. Pilaftsis, Heavy majorana neutrinos and baryogenesis. Int. J. Mod. Phys. A **14**, 1811–1858 (1999). https://doi.org/10.1142/S0217751X99000932
245. T. Asaka, K. Hamaguchi, M. Kawasaki, T. Yanagida, Leptogenesis in inflationary universe. Phys. Rev. D **61**, 083512 (2000). https://doi.org/10.1103/PhysRevD.61.083512
246. W. Buchmuller, M. Plumacher, Neutrino masses and the baryon asymmetry. Int. J. Mod. Phys. A **15**, 5047–5086 (2000). https://doi.org/10.1016/S0217-751X(00)00293-5, https://doi.org/ 10.1142/S0217751X00002935
247. W. Buchmuller, P. Di Bari, M. Plumacher, Leptogenesis for pedestrians. Ann. Phys. **315**, 305–351 (2005). https://doi.org/10.1016/j.aop.2004.02.003
248. W. Buchmuller, R.D. Peccei, T. Yanagida, Leptogenesis as the origin of matter. Ann. Rev. Nucl. Part. Sci. **55**, 311–355 (2005). https://doi.org/10.1146/annurev.nucl.55.090704.151558
249. S. Davidson, E. Nardi, Y. Nir, Leptogenesis. Phys. Rept. **466**, 105–177 (2008). https://doi. org/10.1016/j.physrep.2008.06.002
250. R. Rangarajan, D.V. Nanopoulos. Inflationary baryogenesis. Phys. Rev. D **64**, 063511 (2001). https://doi.org/10.1103/PhysRevD.64.063511
251. S.H.-S. Alexander, M.E. Peskin, M.M. Sheikh-Jabbari, Leptogenesis from gravity waves in models of inflation. Phys. Rev. Lett. **96**, 081301 (2006). https://doi.org/10.1103/PhysRevLett. 96.081301
252. S. Alexander, A. Marciano, D. Spergel, Chern-simons inflation and baryogenesis. JCAP **1304**, 046 (2013). https://doi.org/10.1088/1475-7516/2013/04/046
253. F. Zwicky, Die Rotverschiebung von extragalaktischen Nebeln. Helv. Phys. Acta **6**, 110–127 (1933). https://doi.org/10.1007/s10714-008-0707-4. [Gen. Rel. Grav. 41,207(2009)]
254. V.C. Rubin, W. Kent Ford, Jr., Rotation of the andromeda nebula from a spectroscopic survey of emission regions. Astrophys. J. **159**, 379–403 (1970). https://doi.org/10.1086/150317
255. V.C. Rubin, N. Thonnard, W.K. Ford Jr., Rotational properties of 21 SC galaxies with a large range of luminosities and radii, from NGC 4605 /R = 4kpc/ to UGC 2885 /R = 122 kpc/. Astrophys. J. **238**, 471 (1980). https://doi.org/10.1086/158003
256. V.C. Rubin, D. Burstein, W.K. Ford Jr., N. Thonnard, Rotation velocities of 16 SA galaxies and a comparison of SA, SB, and SC rotation properties. Astrophys. J. **289**, 81 (1985). https:// doi.org/10.1086/162866

257. M. Persic, P. Salucci, Rotation curves of 967 spiral galaxies. Astrophys. J. Suppl. **99**, 501 (1995). https://doi.org/10.1086/192195

258. M. Persic, P. Salucci, F. Stel, The universal rotation curve of spiral galaxies: 1. The dark matter connection. Mon. Not. Roy. Astron. Soc. **281**, 27 (1996). https://doi.org/10.1093/mnras/281.1.27, https://doi.org/10.1093/mnras/278.1.27

259. J.F. Navarro, C.S. Frenk, S.D.M. White, The structure of cold dark matter halos. Astrophys. J. **462**, 563–575 (1996). https://doi.org/10.1086/177173

260. W.J.G. de Blok, F. Walter, E. Brinks, C. Trachternach, S-H. Oh, R.C. Kennicutt, Jr., High-resolution rotation curves and galaxy mass models from THINGS. Astron. J. **136**, 2648–2719 (2008). https://doi.org/10.1088/0004-6256/136/6/2648

261. D.J. Hegyi, K.A. Olive, Can galactic halos be made of baryons? Phys. Lett. B **126**, 28 (1983). https://doi.org/10.1016/0370-2693(83)90009-6

262. D.J. Hegyi, K.A. Olive, A case against baryons in galactic halos. Astrophys. J. **303**, 56–65 (1986). https://doi.org/10.1086/164051

263. G. Steigman, K.A. Olive, D.N. Schramm, Cosmological constraints on superweak particles. Phys. Rev. Lett. **43**, 239–242 (1979). https://doi.org/10.1103/PhysRevLett.43.239

264. K.A. Olive, D.N. Schramm, G. Steigman, Limits on new superweakly interacting particles from primordial nucleosynthesis. Nucl. Phys. B **180**, 497–515 (1981). https://doi.org/10.1016/0550-3213(81)90065-1

265. G. Steigman, M.S. Turner, Cosmological constraints on the properties of weakly interacting massive particles. Nucl. Phys. B **253**, 375–386 (1985). https://doi.org/10.1016/0550-3213(85)90537-1

266. E. Aprile et al., Physics reach of the XENON1T dark matter experiment. JCAP **1604**(04), 027 (2016). https://doi.org/10.1088/1475-7516/2016/04/027

267. D.S. Akerib et al., Results from a search for dark matter in the complete LUX exposure. Phys. Rev. Lett. **118**(2), 021303 (2017). https://doi.org/10.1103/PhysRevLett.118.021303

268. J.R. Primack, D. Seckel, B. Sadoulet, Detection of cosmic dark matter. Ann. Rev. Nucl. Part. Sci. **38**, 751–807 (1988). https://doi.org/10.1146/annurev.ns.38.120188.003535

269. K.A. Olive, TASI lectures on dark matter, in *Particle Physics and Cosmology: The Quest for Physics Beyond the Standard Model(s). Proceedings, Theoretical Advanced Study Institute, TASI 2002, Boulder, USA, June 3–28, 2002* (2003), pp. 797–851

270. L. Bergstrom, Dark matter candidates. New J. Phys. **11**, 105006 (2009). https://doi.org/10.1088/1367-2630/11/10/105006

271. K. Garrett, G. Duda, Dark matter: a primer. Adv. Astron. **2011**, 968283 (2011). https://doi.org/10.1155/2011/968283

272. M. Lisanti, Lectures on dark matter physics, in *Proceedings, Theoretical Advanced Study Institute in Elementary Particle Physics: New Frontiers in Fields and Strings (TASI 2015): Boulder, CO, USA, June 1–26, 2015* (2017), pp. 399–446. https://doi.org/10.1142/9789813149441_0007

273. D.N. Schramm, G. Steigman, Relic neutrinos and the density of the universe. Astrophys. J. **243**, 1 (1981). https://doi.org/10.1086/158559

274. M. Milgrom, A modification of the newtonian dynamics as a possible alternative to the hidden mass hypothesis. Astrophys. J. **270**, 365–370 (1983). https://doi.org/10.1086/161130

275. P. Gondolo, G. Gelmini, Cosmic abundances of stable particles: improved analysis. Nucl. Phys. B **360**, 145–179 (1991). https://doi.org/10.1016/0550-3213(91)90438-4

276. S.M. Barr, Baryogenesis, sphalerons and the cogeneration of dark matter. Phys. Rev. D **44**, 3062–3066 (1991). https://doi.org/10.1103/PhysRevD.44.3062

277. Vadim A. Kuzmin, A simultaneous solution to baryogenesis and dark matter problems. Phys. Part. Nucl. **29**, 257–265 (1998). https://doi.org/10.1134/1.953070. [Phys. Atom. Nucl. 61, 1107 (1998)]

278. D.H. Oaknin, A. Zhitnitsky, Baryon asymmetry, dark matter and quantum chromodynamics. Phys. Rev. D **71**, 023519 (2005). https://doi.org/10.1103/PhysRevD.71.023519

279. R. Kitano, I. Low, Dark matter from baryon asymmetry. Phys. Rev. D **71**, 023510 (2005). https://doi.org/10.1103/PhysRevD.71.023510

280. G.R. Farrar, G. Zaharijas, Dark matter and the baryon asymmetry. Phys. Rev. Lett. **96**, 041302 (2006). https://doi.org/10.1103/PhysRevLett.96.041302

281. D.E. Kaplan, M.A. Luty, K.M. Zurek, Asymmetric dark matter. Phys. Rev. D **79**, 115016 (2009). https://doi.org/10.1103/PhysRevD.79.115016

282. G. Pei-Hong, U. Sarkar, X. Zhang, Visible and dark matter genesis and cosmic positron/electron excesses. Phys. Rev. D **80**, 076003 (2009). https://doi.org/10.1103/PhysRevD.80.076003

283. H. Davoudiasl, D.E. Morrissey, K. Sigurdson, S. Tulin, Hylogenesis: a unified origin for baryonic visible matter and antibaryonic dark matter. Phys. Rev. Lett. **105**, 211304 (2010). https://doi.org/10.1103/PhysRevLett.105.211304

284. N. Haba, S. Matsumoto, Baryogenesis from dark sector. Prog. Theor. Phys. **125**, 1311–1316 (2011). https://doi.org/10.1143/PTP.125.1311

285. G. Pei-Hong, M. Lindner, U. Sarkar, X. Zhang, WIMP dark matter and baryogenesis. Phys. Rev. D **83**, 055008 (2011). https://doi.org/10.1103/PhysRevD.83.055008

286. L.J. Hall, J. March-Russell, S.M. West, A unified theory of matter genesis: asymmetric freeze-in (2010), arXiv:1010.0245

287. B. Dutta, J. Kumar, Asymmetric dark matter from hidden sector baryogenesis. Phys. Lett. B **699**, 364–367 (2011). https://doi.org/10.1016/j.physletb.2011.04.036

288. A. Falkowski, J.T. Ruderman, T. Volansky, Asymmetric dark matter from leptogenesis. JHEP **05**, 106 (2011). https://doi.org/10.1007/JHEP05(2011)106

289. J.J. Heckman, S.-J. Rey, Baryon and dark matter genesis from strongly coupled strings. JHEP **06**, 120 (2011). https://doi.org/10.1007/JHEP06(2011)120

290. M.L. Graesser, I.M. Shoemaker, L. Vecchi, Asymmetric WIMP dark matter. JHEP **10**, 110 (2011). https://doi.org/10.1007/JHEP10(2011)110

291. N.F. Bell, K. Petraki, I.M. Shoemaker, R.R. Volkas, Pangenesis in a baryon-symmetric universe: dark and visible matter via the affleck-dine mechanism. Phys. Rev. D **84**, 123505 (2011). https://doi.org/10.1103/PhysRevD.84.123505

292. C. Cheung, K.M. Zurek, Affleck-dine cogenesis. Phys. Rev. D **84**, 035007 (2011). https://doi.org/10.1103/PhysRevD.84.035007

293. J. March-Russell, M. McCullough, Asymmetric dark matter via spontaneous co-genesis. JCAP **1203**, 019 (2012). https://doi.org/10.1088/1475-7516/2012/03/019

294. Y. Cui, L. Randall, B. Shuve, Emergent dark matter, baryon, and lepton numbers. JHEP **08**, 073 (2011). https://doi.org/10.1007/JHEP08(2011)073

295. A. Mazumdar, The origin of dark matter, matter-anti-matter asymmetry, and inflation (2011), arXiv:1106.5408

296. T. Lin, Y. Hai-Bo, K.M. Zurek, On symmetric and asymmetric light dark matter. Phys. Rev. D **85**, 063503 (2012). https://doi.org/10.1103/PhysRevD.85.063503

297. K. Petraki, M. Trodden, R.R. Volkas, Visible and dark matter from a first-order phase transition in a baryon-symmetric universe. JCAP **1202**, 044 (2012). https://doi.org/10.1088/1475-7516/2012/02/044

298. W.-Z. Feng, A. Mazumdar, P. Nath, Baryogenesis from dark matter. Phys. Rev. D **88**(3), 036014 (2013). https://doi.org/10.1103/PhysRevD.88.036014

299. K.M. Zurek, Asymmetric dark matter: theories, signatures, and constraints. Phys. Rept. **537**, 91–121 (2014). https://doi.org/10.1016/j.physrep.2013.12.001

300. P. Ramond, Neutrinos: a glimpse beyond the standard model. Nucl. Phys. Proc. Suppl. **77**, 3–9 (1999). https://doi.org/10.1016/S0920-5632(99)00382-5

301. Y. Fukuda et al., Measurement of a small atmospheric muon-neutrino / electron-neutrino ratio. Phys. Lett. B **433**, 9–18 (1998). https://doi.org/10.1016/S0370-2693(98)00476-6

302. Y. Fukuda et al., Study of the atmospheric neutrino flux in the multi-GeV energy range. Phys. Lett. B **436**, 33–41 (1998). https://doi.org/10.1016/S0370-2693(98)00876-4

303. Y. Fukuda et al., Evidence for oscillation of atmospheric neutrinos. Phys. Rev. Lett. **81**, 1562–1567 (1998). https://doi.org/10.1103/PhysRevLett.81.1562

304. Y. Fukuda et al., Measurements of the solar neutrino flux from Super-Kamiokande's first 300 days. Phys. Rev. Lett. **81**, 1158–1162 (1998). https://doi.org/10.1103/PhysRevLett.81.1158. [Erratum: Phys. Rev. Lett. 81, 4279 (1998)]

305. Y. Fukuda et al., Constraints on neutrino oscillation parameters from the measurement of day night solar neutrino fluxes at Super-Kamiokande. Phys. Rev. Lett. **82**, 1810–1814 (1999). https://doi.org/10.1103/PhysRevLett.82.1810

306. Y. Fukuda et al., Measurement of the solar neutrino energy spectrum using neutrino electron scattering. Phys. Rev. Lett. **82**, 2430–2434 (1999). https://doi.org/10.1103/PhysRevLett.82.2430

307. Q.R. Ahmad et al., Measurement of the rate of $\nu_e + d \to p + p + e^-$ interactions produced by ^8B solar neutrinos at the Sudbury Neutrino Observatory. Phys. Rev. Lett. **87**, 071301 (2001). https://doi.org/10.1103/PhysRevLett.87.071301

308. Q.R. Ahmad et al., Direct evidence for neutrino flavor transformation from neutral current interactions in the Sudbury Neutrino Observatory. Phys. Rev. Lett. **89**, 011301 (2002). https://doi.org/10.1103/PhysRevLett.89.011301

309. Q.R. Ahmad et al., Measurement of day and night neutrino energy spectra at SNO and constraints on neutrino mixing parameters. Phys. Rev. Lett. **89**, 011302 (2002). https://doi.org/10.1103/PhysRevLett.89.011302

310. B.T. Cleveland, T. Daily, R. Davis, Jr., J.R. Distel, K. Lande, C.K. Lee, P.S. Wildenhain, J. Ullman, Measurement of the solar electron neutrino flux with the Homestake chlorine detector. Astrophys. J. **496**, 505–526 (1998). https://doi.org/10.1086/305343

311. V.N. Gavrin, Solar neutrino results from SAGE. Nucl. Phys. Proc. Suppl. **91**, 36–43 (2001). https://doi.org/10.1016/S0920-5632(00)00920-8

312. E. Bellotti, First results from GNO. Nucl. Phys. Proc. Suppl. **91**, 44–49 (2001). https://doi.org/10.1016/S0920-5632(00)00921-X

313. K.N. Abazajian, Sterile neutrinos in cosmology. Phys. Rept. **711–712**, 1–28 (2017). https://doi.org/10.1016/j.physrep.2017.10.003

314. J. Aasi et al., Advanced LIGO. Class. Quant. Grav. **32**, 074001 (2015). https://doi.org/10.1088/0264-9381/32/7/074001

315. B.P. Abbott et al., Binary black hole mergers in the first advanced LIGO observing run. Phys. Rev. **X6**(4), 041015 (2016). https://doi.org/10.1103/PhysRevX.6.041015

316. B.P. Abbott et al., The rate of binary black hole mergers inferred from advanced LIGO observations surrounding GW150914. Astrophys. J. **833**(1), L1 (2016). https://doi.org/10.3847/2041-8205/833/1/L1

317. B.P. Abbott et al., Observation of gravitational waves from a binary black hole merger. Phys. Rev. Lett. **116**(6), 061102 (2016). https://doi.org/10.1103/PhysRevLett.116.061102

318. B.P. Abbott et al., Properties of the binary black hole merger GW150914. Phys. Rev. Lett. **116**(24), 241102 (2016). https://doi.org/10.1103/PhysRevLett.116.241102

319. B.P. Abbott et al., Tests of general relativity with GW150914. Phys. Rev. Lett. **116**(22), 221101 (2016). https://doi.org/10.1103/PhysRevLett.116.221101

320. B.P. Abbott et al., GW151226: observation of gravitational waves from a 22-Solar-Mass binary black hole coalescence. Phys. Rev. Lett. **116**(24), 241103 (2016). https://doi.org/10.1103/PhysRevLett.116.241103

321. A. Albert et al., Search for high-energy neutrinos from gravitational wave event GW151226 and candidate LVT151012 with ANTARES and icecube. Phys. Rev. D **96**(2), 022005 (2017). https://doi.org/10.1103/PhysRevD.96.022005

322. P. Amaro Seoane et al., The gravitational universe (2013), arXiv:1305.5720

323. C. Caprini et al., Science with the space-based interferometer eLISA. II: gravitational waves from cosmological phase transitions. JCAP **1604**(04), 001 (2016). https://doi.org/10.1088/1475-7516/2016/04/001

324. E. Gildener, Gauge symmetry hierarchies. Phys. Rev. D **14**, 1667 (1976). https://doi.org/10.1103/PhysRevD.14.1667

325. S. Weinberg, Gauge hierarchies. Phys. Lett. B **82**, 387–391 (1979). https://doi.org/10.1016/0370-2693(79)90248-X

326. G. Hooft, Naturalness, chiral symmetry, and spontaneous chiral symmetry breaking. NATO Sci. Ser. B **59**, 135–157 (1980). https://doi.org/10.1007/978-1-4684-7571-5_9

327. E. Gildener, Gauge symmetry hierarchies revisited. Phys. Lett. B **92**, 111–114 (1980). https://doi.org/10.1016/0370-2693(80)90316-0
328. C. Wetterich, Fine tuning problem and the renormalization group. Phys. Lett. B **140**, 215–222 (1984). https://doi.org/10.1016/0370-2693(84)90923-7
329. M.J.G. Veltman, The infrared-ultraviolet connection. Acta Phys. Polon. B **12**, 437 (1981)
330. W.A. Bardeen, On naturalness in the standard model, in *Ontake Summer Institute on Particle Physics Ontake Mountain, Japan, August 27–September 2, 1995* (1995)
331. H. Aoki, S. Iso, Revisiting the naturalness problem-who is afraid of quadratic divergences? Phys. Rev. D **86**, 013001 (2012). https://doi.org/10.1103/PhysRevD.86.013001
332. S.P. Martin, A supersymmetry primer. Adv. Ser. Direct. High Energy Phys. **21**, 1–153 (2010). https://doi.org/10.1142/9789814307505_0001
333. K.A. Meissner, H. Nicolai, Conformal symmetry and the standard model. Phys. Lett. B **648**, 312–317 (2007). https://doi.org/10.1016/j.physletb.2007.03.023
334. Y. Fujii, Scalar-tensor theory of gravitation and spontaneous breakdown of scale invariance. Phys. Rev. D **9**, 874–876 (1974). https://doi.org/10.1103/PhysRevD.9.874
335. F. Englert, C. Truffin, R. Gastmans, Conformal invariance in quantum gravity. Nucl. Phys. B **117**, 407–432 (1976). https://doi.org/10.1016/0550-3213(76)90406-5
336. W. Buchmuller, N. Dragon, Scale invariance and spontaneous symmetry breaking. Phys. Lett. B **195**, 417–422 (1987). https://doi.org/10.1016/0370-2693(87)90041-4
337. C. Wetterich, Cosmology and the fate of dilatation symmetry. Nucl. Phys. B **302**, 668–696 (1988). https://doi.org/10.1016/0550-3213(88)90193-9
338. W. Buchmuller, N. Dragon, Dilatons in flat and curved space-time. Nucl. Phys. B **321**, 207–231 (1989). https://doi.org/10.1016/0550-3213(89)90249-6
339. R. Foot, A. Kobakhidze, R.R. Volkas, Electroweak higgs as a pseudo-goldstone boson of broken scale invariance. Phys. Lett. B **655**, 156–161 (2007). https://doi.org/10.1016/j.physletb.2007.06.084
340. R. Foot, A. Kobakhidze, K.L. McDonald, R.R. Volkas, Neutrino mass in radiatively-broken scale-invariant models. Phys. Rev. D **76**, 075014 (2007). https://doi.org/10.1103/PhysRevD.76.075014
341. R. Foot, A. Kobakhidze, K.L. McDonald, R.R. Volkas, A solution to the hierarchy problem from an almost decoupled hidden sector within a classically scale invariant theory. Phys. Rev. D **77**, 035006 (2008). https://doi.org/10.1103/PhysRevD.77.035006
342. K.A. Meissner, H. Nicolai, Effective action, conformal anomaly and the issue of quadratic divergences. Phys. Lett. B **660**, 260–266 (2008). https://doi.org/10.1016/j.physletb.2007.12.035
343. M. Shaposhnikov, D. Zenhausern, Scale invariance, unimodular gravity and dark energy. Phys. Lett. B **671**, 187–192 (2009). https://doi.org/10.1016/j.physletb.2008.11.054
344. M. Shaposhnikov, D. Zenhausern, Quantum scale invariance, cosmological constant and hierarchy problem. Phys. Lett. B **671**, 162–166 (2009). https://doi.org/10.1016/j.physletb.2008.11.041
345. M.E. Shaposhnikov, F.V. Tkachov, Quantum scale-invariant models as effective field theories (2009), arXiv:0905.4857
346. R. Foot, A. Kobakhidze, R.R. Volkas, Stable mass hierarchies and dark matter from hidden sectors in the scale-invariant standard model. Phys. Rev. D **82**, 035005 (2010). https://doi.org/10.1103/PhysRevD.82.035005
347. R. Foot, A. Kobakhidze, R.R. Volkas, Cosmological constant in scale-invariant theories. Phys. Rev. D **84**, 075010 (2011). https://doi.org/10.1103/PhysRevD.84.075010
348. R. Foot, A. Kobakhidze, Electroweak scale invariant models with small cosmological constant. Int. J. Mod. Phys. A **30**(21), 1550126 (2015). https://doi.org/10.1142/S0217751X15501262
349. D. Blas, M. Shaposhnikov, D. Zenhausern, Scale-invariant alternatives to general relativity. Phys. Rev. D **84**, 044001 (2011). https://doi.org/10.1103/PhysRevD.84.044001
350. I. Oda, Classically scale-invariant B-L model and conformal gravity. Phys. Lett. B **724**, 160–164 (2013). https://doi.org/10.1016/j.physletb.2013.06.014

351. R. Armillis, A. Monin, M. Shaposhnikov, Spontaneously broken conformal symmetry: dealing with the trace anomaly. JHEP **10**, 030 (2013). https://doi.org/10.1007/JHEP10(2013)030
352. I. Bars, P. Steinhardt, N. Turok, Local conformal symmetry in physics and cosmology. Phys. Rev. D **89**(4), 043515 (2014). https://doi.org/10.1103/PhysRevD.89.043515
353. C. Tamarit, Running couplings with a vanishing scale anomaly. JHEP **12**, 098 (2013). https://doi.org/10.1007/JHEP12(2013)098
354. C. Csaki, N. Kaloper, J. Serra, J. Terning, Inflation from broken scale invariance. Phys. Rev. Lett. **113**, 161302 (2014). https://doi.org/10.1103/PhysRevLett.113.161302
355. R.H. Boels, W. Wormsbecher, Spontaneously broken conformal invariance in observables (2015), arXiv:1507.08162
356. P.G. Ferreira, C.T. Hill, G.G. Ross, Scale-independent inflation and hierarchy generation. Phys. Lett. B **763**, 174–178 (2016). https://doi.org/10.1016/j.physletb.2016.10.036
357. G.K. Karananas, M. Shaposhnikov, Scale invariant alternatives to general relativity. II. Dilaton properties. Phys. Rev. D **93**(8), 084052 (2016). https://doi.org/10.1103/PhysRevD.93.084052
358. K. Kannike, M. Raidal, C. Spethmann, H. Veermäe, The evolving Planck mass in classically scale-invariant theories. JHEP **04**, 026 (2017). https://doi.org/10.1007/JHEP04(2017)026
359. G.K. Karananas, J. Rubio, On the geometrical interpretation of scale-invariant models of inflation. Phys. Lett. B **761**, 223–228 (2016). https://doi.org/10.1016/j.physletb.2016.08.037
360. P.G. Ferreira, C.T. Hill, G.G. Ross, Weyl current, scale-invariant inflation and planck scale generation. Phys. Rev. D **95**(4), 043507 (2017). https://doi.org/10.1103/PhysRevD.95.043507
361. D.M. Ghilencea, Z. Lalak, P. Olszewski, Standard model with spontaneously broken quantum scale invariance (2016), arXiv:1612.09120
362. P.G. Ferreira, C.T. Hill, G.G. Ross, No fifth force in a scale invariant universe. Phys. Rev. D **95**(6), 064038 (2017). https://doi.org/10.1103/PhysRevD.95.064038
363. A. Kobakhidze, S. Liang, Standard model with hidden scale invariance and light dilaton (2017), arXiv:1701.04927
364. A. Salvio, Inflationary perturbations in no-scale theories. Eur. Phys. J. C **77**(4), 267 (2017). https://doi.org/10.1140/epjc/s10052-017-4825-6

Chapter 2
Scale Invariant Inflation

The period of inflation in the early universe is a well-established paradigm in standard cosmology due to its success at solving various observational problems, as well as providing measurable predictions. It was first introduced to solve observational problems associated with the standard big bang cosmology, and is now a well-established theory with many models having been proposed and significant effort expended in the pursuit of experimental verification. Although it is generally agreed that there was an epoch of inflation prior to BBN, the exact mechanism that led to the accelerated expansion is still unknown. A problem which makes this more difficult is the degeneracy of the predictions of many models, and the insufficient sensitivity in current measurements of inflationary scenario observables.

The most common approach to inducing this inflationary scenario is the introduction of a scalar field which acts as the inflaton, whose domination of the early universe's energy density, and traversal of its almost flat potential, leads to the exponential expansion. In this chapter we explore a new class of natural inflation mechanisms which exhibit scale invariance via the dilaton, and involve an arbitrary number of scalar fields that are non-minimally coupled to gravity. The scale invariance of the theory assures the flatness of the inflationary potential, which is then lifted by small quantum corrections that violate the conformal symmetry. The breaking of the associated scale symmetry can also provide an origin for the apparent hierarchy of scales observed in nature [1].

2.1 The Inflationary Epoch

Cosmic inflation is an attractive paradigm that resolves some outstanding puzzles of the standard hot Big Bang cosmology, such as the horizon, flatness, and monopole problems [2–14]. In addition, it provides a natural mechanism for the generation of nearly scale-invariant inhomogeneities through the quantum fluctuations of the

© Springer International Publishing AG, part of Springer Nature 2018   43
N. D. Barrie, *Cosmological Implications of Quantum Anomalies*,
Springer Theses, https://doi.org/10.1007/978-3-319-94715-0_2

Fig. 2.1 The CMB as measured by the PLANCK satellite [17]

inflaton field, that at later stages result in the observed large scale structure of the universe [15]. Observations of the CMB and large scale structure provide strong support for cosmic inflation [16] (Fig. 2.1).

Many mechanisms for the inducement of the inflationary epoch have been proposed since the paradigms inception. The majority of these ideas centre on the existence of a scalar field that is homogeneous and isotropic, and dominates the energy density of the universe, thus causing exponential expansion [8, 15, 18–38]. Given the rapid increase in the number of models proposed, work has recently gone into identifying classes of inflationary models [39–42]. This is in part due to the degeneracy of the simplest predictions of observable parameters between different models [43–52].

There are also issues relating to the consistency of the inflationary theory, due to many models requiring the inflaton to take superplanckian values during its evolution [46, 53, 54]. Higher order operators, which are suppressed by the Planck scale, start to contribute significantly for large variations of the inflaton field during inflation. The effective field theory approximation, which favours $|\varphi| \ll M_p$, breaks down in such cases, and inflationary predictions may become unreliable [55, 56].

2.2 Modelling the Inflationary Epoch

The simplest mechanisms of inflation are proposed to be related to the evolution of a scalar field, known as the inflaton field [57–63]. As an illustrative example, we consider a simple version of this mechanism known as chaotic slow roll inflation, which is depicted in Fig. 2.2. This model assumes that at the beginning of the universe the inflaton field is in an unstable vacuum state. Due to the quantum fluctuations of the inflaton field, it will eventually begin to roll down the potential well towards the true vacuum. As the inflaton field rolls slowly down the potential, inflation is occurring;

Fig. 2.2 Illustration of the features of the chaotic slow roll inflationary mechanism [65]

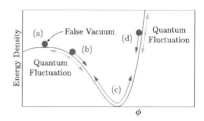

this requires that the field is rolling slower than the rate of expansion. This ensures that as the inflaton rolls down the potential, the potential energy density remains approximately constant leading to a vacuum dominated epoch. Therefore, from the Friedman equations, the scale factor during the inflationary period, expressed in conformal coordinates, is,

$$a(\tau) \propto -\frac{1}{H\tau} \,, \tag{2.1}$$

where $\tau \in [-\infty, 0]$ is the conformal time, with $\tau \to 0$ as inflation proceeds. During this epoch the density of the other forms of energy quickly dilute away. At the end of inflation there is a period known as reheating, prior to the radiation epoch, in which the initial potential energy of the inflaton field is converted into SM particles, producing the matter energy densities we see today [64]. The properties of the reheating epoch will be discussed further in Chap. 4.

We now want to describe the evolution of the inflaton field, so that we can determine the conditions required for inflation to occur, and the possible observable predictions we can obtain from such a mechanism. The inflaton is a scalar field described by the following Lagrangian,

$$\mathcal{L} = -\partial_\mu \phi \partial^\mu \phi - V(\phi), \tag{2.2}$$

where ϕ is the inflaton field, and $V(\phi)$ is the inflationary potential, which could be of the following form,

$$V(\phi) = m^2\phi^2 + \phi^4 \sum_{n=0}^{\infty} \lambda_n \left(\frac{\phi}{M_P}\right)^n . \tag{2.3}$$

We shall now consider the inflaton in a flat FRW universe. In this spacetime background the inflaton has the following equation of motion [66],

$$\ddot{\phi} + 3H\dot{\phi} + V'(\phi) = 0 \,, \tag{2.4}$$

where H is the Hubble parameter which defines the expansion rate, and $V'(\phi)$ is the first derivative with respect to ϕ of the general inflaton potential. This has the corresponding Friedmann equation,

$$H^2 = \frac{3}{M_p^2} \left(\frac{1}{2} \dot{\phi}^2 + V(\phi) \right) , \tag{2.5}$$

where $M_p = 1/\sqrt{8\pi G}$ is the reduced Planck mass, in which G is the gravitational constant. The inflaton field must slowly rolling down its potential for long enough such that the required amount of spatial expansion is achieved to solve the known cosmological problems. In order for this to happen the following conditions need to be satisfied. Firstly, the inflaton's potential energy must dominate the kinetic energy of the field, such that its evolution is gradual, which means,

$$\dot{\phi}^2 \ll V(\phi) , \tag{2.6}$$

The other condition, is that the field's acceleration should be small, allowing a sufficient length of time for slow rolling and hence the inflationary setting to progress. That is,

$$\ddot{\phi} \ll |3H\dot{\phi}|, |V'(\phi)| . \tag{2.7}$$

These requirements allow the definition of the well-known slow-roll parameters, which must be much less than one to ensure the inflaton is slowly rolling and inflation is occurring. The slow roll parameters are as follows,

$$\epsilon = \frac{M_p^2}{2} \left(\frac{V'(\phi)}{V(\phi)} \right)^2 = \frac{1}{2M_P^2} \frac{\dot{\phi}^2}{H^2} , \tag{2.8}$$

$$\eta = M_p^2 \frac{V''(\phi)}{V(\phi)} = -\frac{\ddot{\phi}}{H\dot{\phi}} . \tag{2.9}$$

where these are the leading order slow roll parameters. The ability of a given scenario to successfully lead to an inflationary epoch requires the smallness of the slow-roll parameters, which is equivalent to flatness in the inflaton potential. These constraints ensure that we have an epoch of effective vacuum domination, acting like a cosmological constant. The smaller ϵ and η are, the longer the possible duration of inflation; with the end of inflation, and onset of reheating, occurring when these slow roll parameters are violated. The potential of the ϕ field can now be chosen to see whether it is consistent with an inflationary setting.

During the period of inflation, quantum fluctuations of the inflaton are inflated to produce the temperature anisotropies observed in the CMB today. The level of inhomogeneity produced in the inflationary epoch is related to the properties of the inflationary potential through the slow roll parameters. From the slow roll parameters the power spectrum of the observed scalar perturbations, P_s, the tensor-to-scalar ratio, r, and the spectral index n_s can be determined, and compared with observation. These measurable model predictions are given by,

$$P_s = \frac{1}{24\pi^2 M_p^4}\frac{V(\phi_\star)}{\epsilon_\star} \, , \tag{2.10}$$

$$r = 16\epsilon_\star \, , \tag{2.11}$$

$$n_s = 1 - 6\epsilon_\star + 2\eta_\star \, . \tag{2.12}$$

All the quantities with subscript '$\star$' in the above equations are evaluated at a field value $\phi = \phi_\star$ that corresponds to the field value after a number of e-folds of inflation, $N_\star$. The number of inflationary e-folds can be calculated approximately using the following relation,

$$N_\star \simeq \frac{1}{M_p}\int_0^{\varphi_\star}\frac{d\varphi}{\sqrt{2\epsilon}} \, . \tag{2.13}$$

In this analysis we have considered a single scalar field, but it is also possible to consider inflationary scenarios which contain multiple scalar fields, each of which could potentially contribute to the inflationary expansion [67–80]. In multifield inflation, the number of e-folds is a function of each of the scalar fields in the theory. One scalar may dominate and hence lead to an effective single field inflationary scenario, or you may get a mixture of the scalars contributing to produce a flat direction. Analogously, slow roll parameters and inflationary predictions of a similar form can be derived. In this chapter, we will be considering such a possibility.

The inflationary epoch has also been postulated to solve various other problems in particle physics, this shall be explored in Chap. 3 in relation to Baryogenesis and dark matter. These models can utilise the de Sitter nature of the inflationary setting, the associated reheating epoch, or the inflaton and its decay [81–84]. The possibility of using the reheating epoch for Baryogenesis will be considered in Chap. 4.

A class of natural inflation models has been suggested as a symmetry-motivated solution to the hierarchy problem in particle physics [19]. The inflaton in this class of models is a pseudo-Goldstone boson of a spontaneously broken anomalous global symmetry. The flatness of the inflaton's potential is guaranteed by an approximate shift symmetry in the Lagrangian, meaning that the inflationary potential can easily satisfy the slow roll parameters, and hence can support an inflationary setting. However, the most recent observational results suggest that the simplest models of natural inflation are now disfavoured at 95% confidence level [16]. This type of model is of interest to the work considered in the rest of this chapter.

2.3 Scale Invariance, Weyl Transformations, and Non-minimal Couplings

Before considering the new inflationary mechanism that we will discuss in this chapter, we must first introduce the concept of scale invariance within the context of a scalar Lagrangian including gravity, and the associated Weyl transformation. In some earlier works the existence of a symmetry associated with scale invariance was advocated as a possible explanation for the hierarchy problem without fine-tuning [85, 86]. This idea has received renewed attention in recent times due to the non-observation of Supersymmetry [87, 88]. The existence of a scale invariance symmetry in nature would have many ramifications for particle physics and cosmology. One of the interesting applications of the scale invariance symmetry is to the inflationary scenario, which has been considered more recently in the literature [89–104].

A scale invariant theory is one where there are no physical scales, which is manifest in the classical action through the absence of dimensionful couplings. In this scenario any scales we observe must be dynamically generated, such as through a scalar taking a vacuum expectation value. This would be induced through quantum corrections, which break the scale invariance symmetry explicitly, as the scale invariance symmetry is typically anomalous.

An example of a scale invariant action that includes gravity is,

$$S = \int dx^4 \sqrt{-g} \left(\xi s^2 R - \frac{1}{2} \partial_\mu s \partial^\mu s - V(s) \right) , \qquad (2.14)$$

where we have a SM singlet scalar field s with a scale invariant potential $V(s)$, and ξ is the coupling of s to gravity.

Non-minimal coupling of a scalar field to gravity has been widely discussed in the literature [105, 106]. This idea can lead to many interesting implications for cosmology, particularly for inflation [107–110] and cosmological evolution [111]. The idea of a non-minimal coupling to gravity has been utilised to salvage the idea of Higgs inflation, reconciling the known properties of the Higgs boson with the observational constraints of the associated inflationary scenario. Although such a solution requires the Higgs to have a large non-minimal coupling to gravity, namely $\xi \sim \mathcal{O}(10^5)$ [112].

The Einstein frame, that containing the usual Einstein-Hilbert term, can be transformed to the Jordan frame through the consideration of an extra scalar degree of freedom [113]. In the Jordan frame, the Einstein-Hilbert term is not present, and is replaced by a $\xi s^2 R$ term, as in the action in Eq. (2.14). The coefficient in front of this new term defines the nature of the scalar's coupling to gravity. In the case that $\xi = -1/12$, this is known as the conformal coupling, meaning that the gravitational coupling is of the usual form. When making the Weyl transformation from this frame to the Einstein frame, this scalar is found to not have physical implications and is a fictitious degree of freedom. If instead $\xi \neq -1/12$, then this is known as a non-minimal coupling, and corresponds to an alteration to the gravitational-matter field interactions in the Einstein frame. The advantage of transforming to the Einstein frame from the Jordan frame, in our model, is the added simplicity of inflationary calculations in the Einstein frame.

We shall now exhibit the properties of the Weyl rescaling utilising the action given in Eq. (2.14). This action is expressed in the Jordan frame, in which there is no Einstein-Hilbert term. In order to obtain the Einstein frame we must make a conformal transformation, or Weyl rescaling, which is defined as,

$$g_{\mu\nu}^{E} = \Omega(s)^2 g_{\mu\nu} , \quad \text{where} \quad \Omega(s)^2 = \frac{\xi s^2}{\bar{M}_p^{\,2}} , \tag{2.15}$$

where $g_{\mu\nu}^{E}$ is the metric in the Einstein frame, and $\bar{M}_p$ is the reduced Planck mass. In order for the action to remain invariant under the Weyl rescaling, the scalar and fermion fields must also be transformed as follows,

$$\sigma_E = \frac{\sigma}{\Omega(s)} \quad \text{and} \quad \psi_E = \frac{\psi}{\Omega(s)^{3/2}} , \tag{2.16}$$

while gauge vector fields are invariant under the transformation.

After undertaking this transformation the usual Einstein-Hilbert term is obtained, and the scalar potential is now,

$$V_E(s) = \frac{V_J(s)}{\Omega(s)^4} . \tag{2.17}$$

The scale invariance symmetry can be introduced into a theory through the introduction of a dilaton χ, which is the scalar field associated with the dilatation current. By scaling any coupling with dimension n by $\left(\frac{\chi}{f}\right)^n$ we can remove all the scales in a theory leaving only dimensionless couplings, where f is identified as the dilaton decay constant. The scales we see today would then be generated dynamically by the dilaton when it takes a vacuum expectation value, which breaks the scale invariance symmetry spontaneously. This is how the scale invariance symmetry shall be realised in our inflationary mechanism.

2.4 Natural Inflation with Hidden Scale Invariance

The main focus of this chapter is on a new class of natural inflation models that we have proposed, which are based on a hidden scale invariance. This is realised through the introduction of a spontaneously broken anomalous scaling symmetry with a corresponding pseudo-Goldstone boson, that we identify as the dilaton. We begin by considering a very generic scale-invariant model, with an arbitrary number of scalar fields and all allowed scalar interaction operators in the scalar potential, with dimensionful couplings removed through appropriate rescaling by the dilaton. In this scenario there is found to always exist a direction in the field space that is absolutely flat in the classical limit. This bodes well for successful inflation given the ease with which the slow roll parameters can be satisfied by this model. Inflation proceeds along this direction, while the other fields in this parametrisation reside in their respective (meta)stable minima. Although, the flat direction must have a non-negligible slope such that the inflation epoch is not eternal. Upon quantum corrections, the scale invariance of the model is broken, leading to the flat direction being lifted. Now we

shall have a more detailed discussion of the workings of this model, and derive the
observational predictions it produces.

2.4.1   Description of the Model

Consider a multifield inflationary scenario in which we have N scalar fields, $\{\phi_i\}$
(where $i = 1, 2, \ldots, N$), each with a general non-minimal coupling to gravity,
denoted ξ_i. Now take a Wilsonian effective field theory that describes the SM, or
its extension, that is coupled to gravity at an ultraviolet scale we shall call Λ,

$$S_\Lambda = \int dx^4 \sqrt{-g} \left[\left(\frac{M_p^2}{2} + \sum_{i=1}^{N} \xi_i(\Lambda)\phi_i^2 \right) R - \frac{1}{2} \sum_{i=1}^{N} \partial_\mu \phi_i \partial^\mu \phi_i - V(\phi_i) + \ldots \right], \quad (2.18)$$

where $M_p \approx 2.4 \times 10^{18}$ GeV and we use the mostly positive signature for the metric
tensor. Here we have displayed only the scalar sector, which includes the SM Higgs
boson. The scalar potential $V(\phi_i)$ is a generic polynomial of the scalar fields $\{\phi_i\}$
respecting the relevant symmetries of the theory,

$$V(\phi_i) = \sum_{\{i_n\}} \lambda_{i_1,\ldots,i_n}(\Lambda)\phi_{i_1} \ldots \phi_{i_n}, \quad (2.19)$$

where $\lambda_{i_1,\ldots,i_n}(\Lambda)$ is a coupling of mass dimension $(4 - n)$ defined at the Wilsonian
cut-off Λ, while $\xi_i(\Lambda)$ is a dimensionless non-minimal coupling of the scalar field ϕ_i
to gravity. The scale invariance is explicitly broken in Eq. (2.18) by the ultraviolet cut-
off Λ, the Einstein-Hilbert term $\sim M_p^2 R$ and dimensionful couplings $\lambda_{i_1,\ldots,i_n}$ ($n \neq 4$).

Now we propose that the underlying theory exhibits a scale invariance, which in
the effective low-energy theory is implemented via a (non-linear) pseudo-Goldstone
boson, that is the dilaton χ. A simple way to incorporate the dilaton field χ is to
rescale all of the dimensionful parameters in Eq. (2.18) by the respective powers
of χ/f, where f is the dilaton "decay constant". More specifically, the following
transformations are taken in the action given above,

$$\Lambda \to \Lambda \frac{\chi}{f} \equiv \lambda\chi, \quad M_p^2 \to M_p^2 \left(\frac{\chi}{f} \right)^2 \equiv \xi\chi^2, \quad (2.20)$$

$$\lambda_{i_1,\ldots,i_n}(\Lambda) \to \lambda_{i_1,\ldots,i_n}(\Lambda\chi/f) \left(\frac{\chi}{f} \right)^{4-n} \equiv \sigma_{i_1,\ldots,i_n}(\lambda\chi)\chi^{4-n}. \quad (2.21)$$

Thus, instead of Eq. (2.18), we consider the transformed action,

$$S_{\lambda\chi} = \int dx^4 \sqrt{-g} \left[\left(\xi\chi^2 + \sum_{i=1}^{N} \xi_i(\lambda\chi)\phi_i^2 \right) R \right.$$

$$\left. - \frac{1}{2}\partial_\mu\chi\partial^\mu\chi - \frac{1}{2}\sum_{i=1}^{N}\partial_\mu\phi_i\partial^\mu\phi_i - V(\phi_i, \chi) + \ldots \right], \qquad (2.22)$$

where the scalar potential is now given by,

$$V(\phi_i, \chi) = \sum_{\{i_n\}} \sigma_{i_1,\ldots,i_n}(\lambda\chi) \, \chi^{(4-n)}\phi_{i_1} \ldots \phi_{i_n} \,. \qquad (2.23)$$

This action is manifestly scale invariant in the classical limit. As we shall see, this scale invariance is broken at the quantum level through the renormalisation group (RG) running of the dimensionless couplings, i.e., $\frac{\partial \sigma_{i_1,\ldots,i_n}}{\partial\chi} \neq 0$, etc.

In the calculations that follow, it is convenient to use a 'hyperspherical' representation for the set of scalar fields $\{\phi_i, \chi\}$,

$$\phi_i = \rho\cos(\theta_i)\prod_{k=1}^{i-1}\sin(\theta_k) \,, \quad (i = 1, 2, \ldots, N) \,,$$

$$\chi = \rho\prod_{k=1}^{N}\sin(\theta_k) \,. \qquad (2.24)$$

Expressing the action in Eq. (2.22) through the 'hyperspherical' representation of the fields, we observe that the modulus field ρ factors out. That is, the first term in the action and the scalar potential presented in Eq. (2.22) can be written as $\sim\rho^2\zeta(\theta_i)R$ and $\sim\rho^4 U(\theta_i)$, respectively, in which,

$$\zeta(\theta_i) = \xi(\lambda\chi)\prod_{k=1}^{N}\sin^2(\theta_k) + \sum_{i=1}^{N}\xi_i(\lambda\chi)\cos^2(\theta_i)\prod_{k=1}^{i-1}\sin^2(\theta_k) \,, \qquad (2.25)$$

$$U(\theta_i) = \prod_{k=1}^{N}\sin^{4-n}(\theta_k)\sum_{\{i_n\}}\sigma_{i_1,\ldots,i_n}(\lambda\chi)\cos(\theta_{i_1})\prod_{k=1}^{i_1-1}\sin(\theta_k)\ldots\cos(\theta_{i_n})\prod_{k=1}^{i_n-1}\sin(\theta_k) \qquad (2.26)$$

We further assume that the θ_i fields are relaxed in their stable or sufficiently long-lived (with lifetimes longer than the duration of the observable inflation) minima $\langle\theta_i\rangle = \theta_i^c$ at very early stages in the evolution of the universe. Hence, their dynamics are of no interest to us in what follows, and we can consider the following reduced form of Eq. (2.22),

$$\bar{S}_\rho = \int dx^4 \sqrt{-g} \left[\zeta(\rho)\rho^2 R - \frac{1}{2}\partial_\mu\rho\partial^\mu\rho - V(\rho) \right], \qquad (2.27)$$

$$V(\rho) = \sigma(\rho)\rho^4 \,, \qquad (2.28)$$

where $\zeta \equiv \zeta(\theta_i^c)$ and $\sigma \equiv U(\theta_i^c)$. Hence, we arrive at an effective single-field model with a quartic potential and non-minimal coupling to gravity [114], but without the standard Einstein-Hilbert term. It also resembles the large field limit of the Higgs inflation model [112].

In order to reproduce the Einstein-Hilbert term in Eq. (2.27), the modulus field ρ has to develop a non-zero vacuum expectation value, $\langle \rho \rangle \equiv \rho_0$. If the vacuum configuration $\{\rho_0, \theta_i^c\}$ describes the current vacuum state of the universe, then $\rho_0 = \frac{M_p}{\sqrt{2\zeta(\rho_0)}}$ with $\zeta(\rho_0) \equiv \zeta_0 > 0$. Furthermore, the vacuum energy density, $\frac{\sigma(\rho_0)M_p^4}{4\zeta_0^2}$, in this case must be vanishingly small to satisfy the observational measurements of the dark energy density. That is, the scalar potential must be tuned so that $\sigma(\rho_0) \equiv \sigma_0 \sim 12\zeta_0^2 H_0^2/M_p^2 \approx 0$, where H_0 is the present value of the Hubble parameter. However, inflation may end in a metastable state, which subsequently decays into the current vacuum state, alleviating this possible issue. Hence, we keep ρ_0 and σ_0 as free parameters.

The field ρ represents a flat direction of the potential in Eq. (2.28), which is constant for any value of ρ when considering the Einstein frame. Furthermore, for the special value of the coupling to gravity $\zeta_0 = -1/12$, the conformal coupling, ρ is a fictitious degree of freedom which is not manifest in the Einstein frame. In this case, the action in Eq. (2.27) in fact describes pure Einstein gravity with a cosmological constant.

2.5 Quantum Corrected Potential in Curved Spacetime

The ρ-dependence of the dimensionless couplings σ and ζ in Eqs. (2.27) and (2.28) is determined by computing the quantum-corrected effective potential. At the classical level these couplings are constant and independent of ρ, and hence the action in Eq. (2.27) is classically scale invariant. In our analysis we wish to consider the 1-loop quantum corrections to the scalar potential of ρ, which are generated by the ρ^2 and ρ^4 interaction terms. In the case of these two interaction terms, the possible interactions induced via the radiative corrections will also be of the form ρ^2 and ρ^4 at 1-loop, but also with a logarithmic dependence on ρ analogous to the Colemann Weinberg potential. These corrections can be interpreted as contributions to the running of the two dimensionless couplings σ and ζ.

Seeing as we wish to consider the inflationary setting we calculate the correction to the scalar potential in an FRW spacetime background. We use the closed form effective potential computed in Ref. [115] to obtain the 1-loop approximation of the couplings ζ and σ. The running of the couplings ζ and σ induced by the corrections is described by the β functions, β_σ and β_ζ, which encapsulate the dependence of the couplings on the energy scale. The loop corrected couplings are given by the following,

$$\zeta(\rho) = \zeta_0 + \frac{1}{2}\beta_\zeta \ln\left(\frac{\rho^2}{\rho_0^2}\right) \quad \text{and} \quad \sigma(\rho) = \sigma_0 + \frac{1}{2}\beta_\sigma \ln\left(\frac{\rho^2}{\rho_0^2}\right). \quad (2.29)$$

In our model the corresponding β functions are given by,

$$\frac{\beta_\zeta}{\zeta} = \frac{1}{4\pi^2}\frac{\sigma}{\zeta}(1 + 12\zeta) \quad \text{and} \quad \frac{\beta_\sigma}{\sigma} = \frac{9}{\pi^2}\sigma \; . \tag{2.30}$$

Thus, the 1-loop approximations of the ζ and σ couplings are,

$$\zeta(\rho) = \zeta_0 + \frac{(12\zeta_0 + 1)\sigma_0}{16\pi^2}\ln\left(\frac{\rho^2}{\rho_0^2}\right) , \tag{2.31}$$

$$\sigma(\rho) = \sigma_0 + \frac{9\sigma_0^2}{4\pi^2}\ln\left(\frac{\rho^2}{\rho_0^2}\right) . \tag{2.32}$$

It is clear to see that the classical scale invariance of the theory is broken by these radiative corrections, which is illustrated by the ρ dependence of the couplings found in the 1-loop corrections in Eqs. (2.31) and (2.32). Note that $\sigma_0 \to 0$ is a conformal fixed-point of the theory, since the ρ dependence disappears in Eqs. (2.31) and (2.32) in this limit. The conformal coupling $\zeta_0 = -1/12$ is also a fixed-point as $\zeta(\rho) = \zeta_0$. Hence, having σ small or ζ close to $-1/12$, near the respective fixed points, is natural in the technical sense. All these attractive features motivate us to consider scale invariance as an essential symmetry for natural inflation, with ρ being the inflaton field.

2.6 Observational Signatures and Model Predictions

To compute the inflationary observables we first wish to take the action in Eq. (2.27) and perform a Weyl rescaling to obtain the action in the Einstein frame. The rescaling is as follows,

$$g_{\mu\nu} \to \Omega^2 g_{\mu\nu} , \quad \Omega^2 = \frac{2\zeta\rho^2}{M_p^2} \; . \tag{2.33}$$

We also must make a field redefinition of the inflaton field ρ in order to obtain the canonical form of the kinetic term. The field redefinition is,

$$\rho = \rho_0 \exp\left(\frac{\sqrt{\tilde{\zeta}}}{M_p}\varphi\right) , \tag{2.34}$$

where $\tilde{\zeta} = \frac{2\zeta}{1+12\zeta}$ with $\zeta > 0$ or $\zeta < -1/12$. Therefore, in the Einstein frame the action in Eq. (2.27) reads,

$$\bar{S}_\varphi = \int dx^4 \sqrt{-g} \left[\frac{M_p^2}{2} R - \frac{1}{2} \partial_\mu \varphi \partial^\mu \varphi - V(\varphi) \right], \tag{2.35}$$

$$V(\rho(\varphi)) = \frac{M_p^4}{4} \frac{\sigma(\rho(\varphi))}{\zeta^2(\rho(\varphi))}. \tag{2.36}$$

In order to proceed with the actual calculations of the above observables, we substitute Eqs. (2.31) and (2.32) into Eq. (2.36), and using Eq. (2.34) we can then express the effective potential in terms of the inflaton field φ in the Einstein frame. Next, let us consider the conformal limit where $\sigma_0 \to 0$ and $\zeta_0 \to -1/12$. The latter limit implies that ζ evolves slowly, $\zeta \approx \zeta_0$. If we assume further that $\sigma_0^2 \sqrt{\frac{2\zeta_0}{1+12\zeta_0}}$ approaches to some constant C, the potential in Eq. (2.36) can be well approximated by a potential which is linear in the inflaton field φ. It should be noted, a linear potential was obtained in another limit of the non-minimally coupling in Ref. [116]. Therefore, in our scenario the potential takes the following form,

$$V(\varphi) \approx \frac{162C}{\pi^2} M_p^3 \varphi. \tag{2.37}$$

This linear potential can now be used to compute the slow roll parameters and inflationary observables given in Eqs. (2.8–2.13). These immediately imply that $\eta = 0$ and hence we find the relation for spectral index to be,

$$n_s \approx 1 - \frac{3}{8} r. \tag{2.38}$$

In terms of the number of observable e-folds $N_\star$, the model predictions are,

$$n_s - 1 \approx -0.025 \left(\frac{N_\star}{60} \right)^{-1}, \tag{2.39}$$

$$r = 0.0667 \left(\frac{N_\star}{60} \right)^{-1}. \tag{2.40}$$

If we take the observed value of the scalar perturbations ($P_s \simeq 10^{-9}$) we require that $C \approx 5.5 \cdot 10^{-12} \left(\frac{N_\star}{60} \right)^{-3/2}$. The predictions found in Eq. (2.39) are in reasonable agreement with the most recent analysis of the cosmological data [117], which suggest that,

$$n_s = 0.9669 \pm 0040 \quad (68\%\text{C.L.}), \tag{2.41}$$

$$r_{0.01} < 0.0685 \quad (95\%\text{C.L.}), \tag{2.42}$$

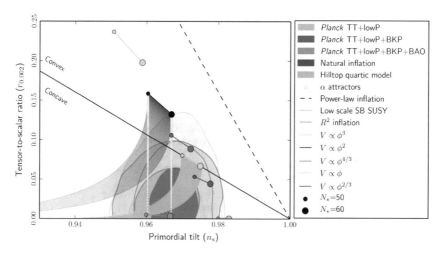

Fig. 2.3 Constraints on Inflationary observables from the PLANCK satellite, including predictions of various inflationary potentials [16]

for the ΛCDM+r model. As can also be seen in Fig. 2.3, the predicted values fit well with observation. Further improvement of the accuracy of cosmological measurements will be critical for this scenario.

Note that, for large ($\xi \to \infty$) and small ($\xi \to 0$) non-minimal couplings, $n_s \gtrsim 1$, and thus the model is excluded by observation in these limiting cases.

2.7 Conclusions and Future Prospects

In this chapter, we have explored a new class of natural inflation models that exhibit hidden scale invariance, which is realised through the dilaton field. A very generic Wilsonian potential was considered, consisting of an arbitrary number of scalar fields, that was found to contain a flat direction in the classical limit, which was lifted by quantum corrections. Thus inflation can naturally, without fine-tuning, proceed as the inflaton field evolves along this direction. We find that in the conformal coupling limit, within the leading perturbative approximation, the generic model is reduced to a one-field model with a linear potential, $V(\varphi) \sim \varphi$, with the linear term being radiatively induced. Such a scenario leads to the specific predictions of the spectral index and the tensor-to-scalar ratio: $n_s - 1 \approx -0.025 \left(\frac{N_*}{60}\right)^{-1}$ and $r \approx 0.0667 \left(\frac{N_*}{60}\right)^{-1}$, respectively. These predictions are in reasonable agreement with observation, but more accurate cosmological measurements are required.

References

1. N.D. Barrie, A. Kobakhidze, S. Liang, Natural inflation with hidden scale invariance. Phys. Lett. B **756**, 390–393 (2016a). https://doi.org/10.1016/j.physletb.2016.03.056
2. A.H. Guth, The inflationary universe: a possible solution to the horizon and flatness problems. Phys. Rev. D **23**, 347–356 (1981). https://doi.org/10.1103/PhysRevD.23.347
3. A.A. Starobinsky, A new type of isotropic cosmological models without singularity. Phys. Lett. B **91**, 99–102 (1980). https://doi.org/10.1016/0370-2693(80)90670-X
4. D. Kazanas, Dynamics of the universe and spontaneous symmetry breaking. Astrophys. J. **241**, L59–L63 (1980). https://doi.org/10.1086/183361
5. K. Sato, First order phase transition of a vacuum and expansion of the universe. Mon. Not. Roy. Astron. Soc. **195**, 467–479 (1981a)
6. A.D. Linde, A new inflationary universe scenario: a possible solution of the horizon, flatness, homogeneity, isotropy and primordial monopole problems. Phys. Lett. B **108**, 389–393 (1982a). https://doi.org/10.1016/0370-2693(82)91219-9
7. A.D. Linde, Coleman-weinberg theory and a new inflationary universe scenario. Phys. Lett. B **114**, 431–435 (1982b). https://doi.org/10.1016/0370-2693(82)90086-7
8. A. Albrecht, P.J. Steinhardt, Cosmology for grand unified theories with radiatively induced symmetry breaking. Phys. Rev. Lett. **48**, 1220–1223 (1982). https://doi.org/10.1103/PhysRevLett.48.1220
9. A.H. Guth, S.Y. Pi, Fluctuations in the new inflationary universe. Phys. Rev. Lett. **49**, 1110–1113 (1982). https://doi.org/10.1103/PhysRevLett.49.1110
10. M.S. Turner, Coherent scalar field oscillations in an expanding universe. Phys. Rev. D **28**, 1243 (1983). https://doi.org/10.1103/PhysRevD.28.1243
11. J.M. Bardeen, P.J. Steinhardt, M.S. Turner, Spontaneous creation of almost scale-free density perturbations in an inflationary universe. Phys. Rev. D **28**, 679 (1983). https://doi.org/10.1103/PhysRevD.28.679
12. A.D. Linde, The inflationary universe. Rept. Prog. Phys. **47**, 925–986 (1984). https://doi.org/10.1088/0034-4885/47/8/002
13. R.H. Brandenberger, Quantum field theory methods and inflationary universe models. Rev. Mod. Phys. **57**, 1 (1985). https://doi.org/10.1103/RevModPhys.57.1
14. A.D. Linde, Eternal chaotic inflation. Mod. Phys. Lett. **A1**, 81 (1986). https://doi.org/10.1142/S0217732386000129
15. V.F. Mukhanov, G.V. Chibisov, Quantum fluctuations and a nonsingular universe. JETP Lett. **33**, 532–535 (1981). [Pisma Zh. Eksp. Teor. Fiz. 33, 549 (1981)]
16. P.A.R. Ade et al., Planck 2015 results. XX. Constraints on inflation. Astron. Astrophys. **594**, A20 (2016b). https://doi.org/10.1051/0004-6361/201525898
17. P.A.R. Ade et al., Planck 2013 results. I. Overview of products and scientific results. Astron. Astrophys. **571**, A1 (2014a). https://doi.org/10.1051/0004-6361/201321529
18. F. Lucchin, S. Matarrese, Power law inflation. Phys. Rev. D **32**, 1316 (1985). https://doi.org/10.1103/PhysRevD.32.1316
19. K. Freese, J.A. Frieman, A.V. Olinto, Natural inflation with pseudo-Nambu-Goldstone bosons. Phys. Rev. Lett. **65**, 3233–3236 (1990). https://doi.org/10.1103/PhysRevLett.65.3233
20. J.D. Barrow, K. Maeda, Extended inflationary universes. Nucl. Phys. B **341**, 294–308 (1990). https://doi.org/10.1016/0550-3213(90)90272-F
21. A.L. Berkin, K.-I. Maeda, Inflation in generalized Einstein theories. Phys. Rev. D **44**, 1691–1704 (1991). https://doi.org/10.1103/PhysRevD.44.1691
22. A.D. Linde, Axions in inflationary cosmology. Phys. Lett. B **259**, 38–47 (1991). https://doi.org/10.1016/0370-2693(91)90130-I
23. L.F. Abbott, E. Farhi, M.B. Wise, Particle production in the new inflationary cosmology. Phys. Lett. B **117**, 29 (1982). https://doi.org/10.1016/0370-2693(82)90867-X
24. S.W. Hawking, The development of irregularities in a single bubble inflationary universe. Phys. Lett. B **115**, 295 (1982). https://doi.org/10.1016/0370-2693(82)90373-2

25. A.A. Starobinsky, Dynamics of phase transition in the new inflationary universe scenario and generation of perturbations. Phys. Lett. B **117**, 175–178 (1982). https://doi.org/10.1016/0370-2693(82)90541-X

26. F.C. Adams, J.R. Bond, K. Freese, J.A. Frieman, A.V. Olinto, Natural inflation: particle physics models, power law spectra for large scale structure, and constraints from COBE. Phys. Rev. D **47**, 426–455 (1993). https://doi.org/10.1103/PhysRevD.47.426

27. A.D. Linde, Chaotic inflation. Phys. Lett. B **129**, 177–181 (1983). https://doi.org/10.1016/0370-2693(83)90837-7

28. A.D. Linde, Hybrid inflation. Phys. Rev. D **49**, 748–754 (1994). https://doi.org/10.1103/PhysRevD.49.748

29. E.J. Copeland, A.R. Liddle, D.H. Lyth, E.D. Stewart, D. Wands, False vacuum inflation with Einstein gravity. Phys. Rev. D **49**, 6410–6433 (1994). https://doi.org/10.1103/PhysRevD.49.6410

30. A. Berera, Warm inflation. Phys. Rev. Lett. **75**, 3218–3221 (1995). https://doi.org/10.1103/PhysRevLett.75.3218

31. P. Binetruy, G.R. Dvali, D term inflation. Phys. Lett. B **388**, 241–246 (1996). https://doi.org/10.1016/S0370-2693(96)01083-0

32. G.R. Dvali, S.H. Henry Tye, Brane inflation. Phys. Lett. B **450**, 72–82 (1999). https://doi.org/10.1016/S0370-2693(99)00132-X

33. D.H. Lyth, A. Riotto, Particle physics models of inflation and the cosmological density perturbation. Phys. Rept. **314**, 1–146 (1999). https://doi.org/10.1016/S0370-1573(98)00128-8

34. A.R. Liddle, A. Mazumdar, F.E. Schunck, Assisted inflation. Phys. Rev. D **58**, 061301 (1998). https://doi.org/10.1103/PhysRevD.58.061301

35. C. Armendariz-Picon, T. Damour, V.F. Mukhanov, k-inflation. Phys. Lett. B **458**, 209–218 (1999). https://doi.org/10.1016/S0370-2693(99)00603-6

36. A. Mazumdar, Extra dimensions and inflation. Phys. Lett. B **469**, 55–60 (1999). https://doi.org/10.1016/S0370-2693(99)01256-3

37. L. Boubekeur, D.H. Lyth, Hilltop inflation. JCAP **0507**, 010 (2005). https://doi.org/10.1088/1475-7516/2005/07/010

38. S. Dimopoulos, S. Kachru, J. McGreevy, J.G. Wacker, N-flation. JCAP **0808**, 003 (2008). https://doi.org/10.1088/1475-7516/2008/08/003

39. D. Roest, Universality classes of inflation. JCAP **1401**, 007 (2014). https://doi.org/10.1088/1475-7516/2014/01/007

40. M. Galante, R. Kallosh, A. Linde, D. Roest, Unity of cosmological inflation attractors. Phys. Rev. Lett. **114**(14), 141302 (2015). https://doi.org/10.1103/PhysRevLett.114.141302

41. P. Binetruy, E. Kiritsis, J. Mabillard, M. Pieroni, C. Rosset, Universality classes for models of inflation. JCAP **1504**(04), 033 (2015). https://doi.org/10.1088/1475-7516/2015/04/033

42. V. Domcke, M. Pieroni, P. Bintruy, Primordial gravitational waves for universality classes of pseudoscalar inflation. JCAP **1606**, 031 (2016). https://doi.org/10.1088/1475-7516/2016/06/031

43. L.F. Abbott, B. Mark, M.B. Wise, Constraints on generalized inflationary cosmologies. Nucl. Phys. B **244**, 541–548 (1984). https://doi.org/10.1016/0550-3213(84)90329-8

44. P.J.E. Peebles, Tests of cosmological models constrained by inflation. Astrophys. J. **284**, 439–444 (1984). https://doi.org/10.1086/162425

45. F.C. Adams, K. Freese, A.H. Guth, Constraints on the scalar field potential in inflationary models. Phys. Rev. D **43**, 965–976 (1991). https://doi.org/10.1103/PhysRevD.43.965

46. D.H. Lyth, What would we learn by detecting a gravitational wave signal in the cosmic microwave background anisotropy? Phys. Rev. Lett. **78**, 1861–1863 (1997). https://doi.org/10.1103/PhysRevLett.78.1861

47. D.N. Spergel, M. Zaldarriaga, CMB polarization as a direct test of inflation. Phys. Rev. Lett. **79**, 2180–2183 (1997). https://doi.org/10.1103/PhysRevLett.79.2180

48. A.R. Liddle, S.M. Leach, How long before the end of inflation were observable perturbations produced? Phys. Rev. D **68**, 103503 (2003). https://doi.org/10.1103/PhysRevD.68.103503

49. L. Alabidi, D.H. Lyth, Inflation models and observation. JCAP **0605**, 016 (2006). https://doi.org/10.1088/1475-7516/2006/05/016

50. J.L. Cook, L. Sorbo, Particle production during inflation and gravitational waves detectable by ground-based interferometers. Phys. Rev. D **85**, 023534 (2012). https://doi.org/10.1103/PhysRevD.86.069901, https://doi.org/10.1103/PhysRevD.85.023534. [Erratum: Phys. Rev. D **86**, 069901 (2012)]

51. D.H. Lyth, The CMB modulation from inflation. JCAP **1308**, 007 (2013). https://doi.org/10.1088/1475-7516/2013/08/007

52. J. Martin, C. Ringeval, R. Trotta, V. Vennin, The best inflationary models after Planck. JCAP **1403**, 039 (2014b). https://doi.org/10.1088/1475-7516/2014/03/039

53. R.H. Brandenberger, J. Martin, Trans-Planckian issues for inflationary cosmology. Class. Quant. Grav. **30**, 113001 (2013). https://doi.org/10.1088/0264-9381/30/11/113001

54. G. Barenboim, O. Vives, Transplanckian masses in inflation. Nucl. Part. Phys. Proc. **273–275**, 446–451 (2016). https://doi.org/10.1016/j.nuclphysbps.2015.09.065

55. C. Cheung, P. Creminelli, A.L. Fitzpatrick, J. Kaplan, L. Senatore, The effective field theory of inflation. JHEP **03**, 014 (2008). https://doi.org/10.1088/1126-6708/2008/03/014

56. S. Weinberg, Effective field theory for inflation. Phys. Rev. D **77**, 123541 (2008a). https://doi.org/10.1103/PhysRevD.77.123541

57. K.A. Olive, Inflation. Phys. Rept. **190**, 307–403 (1990). https://doi.org/10.1016/0370-1573(90)90144-Q

58. E.W. Kolb, M.S. Turner, The early universe. Front. Phys. **69**, 1–547 (1990)

59. A. Riotto, Inflation and the theory of cosmological perturbations, in *Astroparticle Physics and Cosmology. Proceedings: Summer School*, Trieste, Italy, 17 Jun–5 Jul 2002 (2002), pp. 317–413

60. R.H. Brandenberger, Lectures on the theory of cosmological perturbations. Lect. Notes Phys. **646**, 127–167 (2004)

61. A.D. Linde, Particle physics and inflationary cosmology. Contemp. Concepts Phys. **5**, 1–362 (1990)

62. K.A. Malik, D. Wands, Cosmological perturbations. Phys. Rept. **475**, 1–51 (2009). https://doi.org/10.1016/j.physrep.2009.03.001

63. D. Baumann, Inflation, in *Physics of the Large and the Small, TASI 09, Proceedings of the Theoretical Advanced Study Institute in Elementary Particle Physics*, Boulder, Colorado, USA, 1–26 June 2009 (2011), pp. 523–686. https://doi.org/10.1142/9789814327183_0010

64. L. Kofman, A.D. Linde, A.A. Starobinsky, Reheating after inflation. Phys. Rev. Lett. **73**, 3195–3198 (1994). https://doi.org/10.1103/PhysRevLett.73.3195

65. A.H. Guth, D.I. Kaiser, Inflationary cosmology: exploring the universe from the smallest to the largest scales. Science **307**, 884–890 (2005). https://doi.org/10.1126/science.1107483

66. S.M. Carroll, *Spacetime and Geometry: An Introduction to General Relativity* (Addison-Wesley, San Francisco, USA, 2004). ISBN 0805387323, 9780805387322

67. A.A. Starobinsky, Multicomponent de sitter (Inflationary) stages and the generation of perturbations. JETP Lett. **42**, 152–155 (1985)

68. L.A. Kofman, D.Y. Pogosian, Nonflat perturbations in inflationary cosmology. Phys. Lett. B **214**, 508–514 (1988). https://doi.org/10.1016/0370-2693(88)90109-8

69. J. Garcia-Bellido, D. Wands, Metric perturbations in two field inflation. Phys. Rev. D **53**, 5437–5445 (1996). https://doi.org/10.1103/PhysRevD.53.5437

70. L.E. Allen, S. Gupta, D. Wands, Non-Gaussian perturbations from multi-field inflation. JCAP **0601**, 006 (2006). https://doi.org/10.1088/1475-7516/2006/01/006

71. F. Vernizzi, D. Wands, Non-Gaussianities in two-field inflation. JCAP **0605**, 019 (2006). https://doi.org/10.1088/1475-7516/2006/05/019

72. T. Battefeld, R. Easther, Non-Gaussianities in multi-field inflation. JCAP **0703**, 020 (2007). https://doi.org/10.1088/1475-7516/2007/03/020

73. S. Yokoyama, T. Suyama, T. Tanaka, Primordial non-Gaussianity in multi-scalar inflation. Phys. Rev. D **77**, 083511 (2008). https://doi.org/10.1103/PhysRevD.77.083511

74. S. Yokoyama, T. Suyama, T. Tanaka, Primordial non-Gaussianity in multi-scalar slow-roll inflation. JCAP **0707**, 013 (2007). https://doi.org/10.1088/1475-7516/2007/07/013

75. D. Wands, Multiple field inflation. Lect. Notes Phys. **738**, 275–304 (2008). https://doi.org/10.1007/978-3-540-74353-8_8

76. D. Langlois, Cosmological perturbations from multi-field inflation. J. Phys. Conf. Ser. **140**, 012004 (2008). https://doi.org/10.1088/1742-6596/140/1/012004

77. K.-Y. Choi, J.-O. Gong, D. Jeong, Evolution of the curvature perturbation during and after multi-field inflation. JCAP **0902**, 032 (2009). https://doi.org/10.1088/1475-7516/2009/02/032

78. T. Chiba, M. Yamaguchi, Extended slow-roll conditions and primordial fluctuations: multiple scalar fields and generalized gravity. JCAP **0901**, 019 (2009). https://doi.org/10.1088/1475-7516/2009/01/019

79. D.I. Kaiser, Conformal transformations with multiple scalar fields. Phys. Rev. D **81**, 084044 (2010). https://doi.org/10.1103/PhysRevD.81.084044

80. L. Senatore, M. Zaldarriaga, The effective field theory of multifield inflation. JHEP **04**, 024 (2012). https://doi.org/10.1007/JHEP04(2012)024

81. P. Adshead, M. Wyman, Chromo-natural inflation: natural inflation on a steep potential with classical non-abelian gauge fields. Phys. Rev. Lett. **108**, 261302 (2012). https://doi.org/10.1103/PhysRevLett.108.261302

82. M.M. Sheikh-Jabbari, Gauge-flation versus chromo-natural Inflation. Phys. Lett. B **717**, 6–9 (2012). https://doi.org/10.1016/j.physletb.2012.09.014

83. P. Adshead, E. Martinec, M. Wyman, Perturbations in chromo-natural inflation. JHEP **1309**, 087 (2013). https://doi.org/10.1007/JHEP09(2013)087

84. M.P. Hertzberg, J. Karouby, Generating the observed baryon asymmetry from the inflaton field. Phys. Rev. D **89**(6), 063523 (2014). https://doi.org/10.1103/PhysRevD.89.063523

85. C. Wetterich, Fine tuning problem and the renormalization group. Phys. Lett. B **140**, 215–222 (1984). https://doi.org/10.1016/0370-2693(84)90923-7

86. W.A. Bardeen, On naturalness in the standard model, in *Ontake Summer Institute on Particle Physics*, Ontake Mountain, Japan, 27 August– 2 September 1995 (1995)

87. R. Foot, A. Kobakhidze, K.L. McDonald, R.R. Volkas, Poincar protection for a natural electroweak scale. Phys. Rev. D **89**(11), 115018 (2014). https://doi.org/10.1103/PhysRevD.89.115018

88. A. Kobakhidze, K.L. McDonald, Comments on the hierarchy problem in effective theories. JHEP **07**, 155 (2014). https://doi.org/10.1007/JHEP07(2014)155

89. J. Garcia-Bellido, J. Rubio, M. Shaposhnikov, D. Zenhausern, Higgs-dilaton cosmology: from the early to the late universe. Phys. Rev. D **84**, 123504 (2011). https://doi.org/10.1103/PhysRevD.84.123504

90. R. Kallosh, A. Linde, Universality class in conformal inflation. JCAP **1307**, 002 (2013). https://doi.org/10.1088/1475-7516/2013/07/002

91. A. Salvio, A. Strumia, Agravity. JHEP **06**, 080 (2014). https://doi.org/10.1007/JHEP06(2014)080

92. J. Ellis, M.A.G. Garcia, D.V. Nanopoulos, K.A. Olive, A no-scale inflationary model to fit them all. JCAP **1408**, 044 (2014). https://doi.org/10.1088/1475-7516/2014/08/044

93. K. Kannike, A. Racioppi, M. Raidal, Embedding inflation into the Standard Model-more evidence for classical scale invariance. JHEP **06**, 154 (2014). https://doi.org/10.1007/JHEP06(2014)154

94. C. Csaki, N. Kaloper, J. Serra, J. Terning, Inflation from broken scale invariance. Phys. Rev. Lett. **113**, 161302 (2014). https://doi.org/10.1103/PhysRevLett.113.161302

95. K. Kannike, G. Htsi, L. Pizza, A. Racioppi, M. Raidal, A. Salvio, A. Strumia, Dynamically induced Planck scale and inflation. JHEP **05**, 065 (2015a). https://doi.org/10.1007/JHEP05(2015)065

96. M. Ozkan, D. Roest, *Universality Classes of Scale Invariant Inflation* (2015), arXiv:1507.03603

97. K. Kannike, G. Htsi, L. Pizza, A. Racioppi, M. Raidal, A. Salvio, A. Strumia, Dynamically induced Planck scale and inflation. PoS, EPS-HEP2015 **379** (2015b)
98. A. Farzinnia, S. Kouwn, Classically scale invariant inflation, supermassive WIMPs, and adimensional gravity. Phys. Rev. D **93**(6), 063528 (2016). https://doi.org/10.1103/PhysRevD. 93.063528
99. M. Rinaldi, L. Vanzo, Inflation and reheating in theories with spontaneous scale invariance symmetry breaking. Phys. Rev. D **94**(2), 024009 (2016). https://doi.org/10.1103/PhysRevD. 94.024009
100. P.G. Ferreira, C.T. Hill, G.G. Ross, Scale-independent inflation and hierarchy generation. Phys. Lett. B **763**, 174–178 (2016). https://doi.org/10.1016/j.physletb.2016.10.036
101. K. Kannike, M. Raidal, C. Spethmann, H. Veerme, The evolving Planck mass in classically scale-invariant theories. JHEP **04**, 026 (2017). https://doi.org/10.1007/JHEP04(2017)026
102. G.K. Karananas, J. Rubio, On the geometrical interpretation of scale-invariant models of inflation. Phys. Lett. B **761**, 223–228 (2016). https://doi.org/10.1016/j.physletb.2016.08.037
103. P.G. Ferreira, C.T. Hill, G.G. Ross, Weyl current, scale-invariant inflation and Planck scale generation. Phys. Rev. D **95**(4), 043507 (2017a). https://doi.org/10.1103/PhysRevD.95.043507
104. A. Salvio, Inflationary perturbations in no-scale theories. Eur. Phys. J. C **77**(4), 267 (2017). https://doi.org/10.1140/epjc/s10052-017-4825-6
105. S. Sonego, V. Faraoni, Coupling to the curvature for a scalar field from the equivalence principle. Class. Quant. Grav. **10**, 1185–1187 (1993). https://doi.org/10.1088/0264-9381/10/ 6/015
106. A.A. Grib, E.A. Poberii, On the difference between conformal and minimal couplings in general relativity. Helv. Phys. Acta **68**, 380–395 (1995)
107. V. Faraoni, Nonminimal coupling of the scalar field and inflation. Phys. Rev. D **53**, 6813–6821 (1996). https://doi.org/10.1103/PhysRevD.53.6813
108. M.P. Hertzberg, On inflation with non-minimal coupling. JHEP **11**, 023 (2010). https://doi. org/10.1007/JHEP11(2010)023
109. M. Pieroni, β-function formalism for inflationary models with a non minimal coupling with gravity. JCAP **1602**(02), 012 (2016). https://doi.org/10.1088/1475-7516/2016/02/012
110. M. Artymowski, A. Racioppi, Scalar-tensor linear inflation. JCAP **1704**(04), 007 (2017). https://doi.org/10.1088/1475-7516/2017/04/007
111. O. Hrycyna, What ξ? Cosmological constraints on the non-minimal coupling constant. Phys. Lett. B **768**, 218–227 (2017). https://doi.org/10.1016/j.physletb.2017.02.062
112. F.L. Bezrukov, M. Shaposhnikov, The Standard Model Higgs boson as the inflaton. Phys. Lett. B **659**, 703–706 (2008). https://doi.org/10.1016/j.physletb.2007.11.072
113. A.Y. Kamenshchik, C.F. Steinwachs, Question of quantum equivalence between Jordan frame and Einstein frame. Phys. Rev. D **91**(8), 084033 (2015). https://doi.org/10.1103/PhysRevD. 91.084033
114. N. Okada, M.U. Rehman, Q. Shafi, Tensor to scalar ratio in non-minimal ϕ^4 inflation. Phys. Rev. D **82**, 043502 (2010). https://doi.org/10.1103/PhysRevD.82.043502
115. S.D. Odintsov, Two loop effective potential in quantum field theory in curved space-time. Phys. Lett. B **306**, 233–236 (1993). https://doi.org/10.1016/0370-2693(93)90073-Q
116. K. Kannike, A. Racioppi, M. Raidal, Linear inflation from quartic potential. JHEP **01**, 035 (2016). https://doi.org/10.1007/JHEP01(2016)035
117. Q.-G. Huang, K. Wang, S. Wang, Inflation model constraints from data released in 2015. Phys. Rev. D **93**(10), 103516 (2016). https://doi.org/10.1103/PhysRevD.93.103516

Chapter 3
An Asymmetric Universe from Inflation

It is generally considered that the generation of the observed matter-antimatter asymmetry must have occurred after the inflationary epoch, as otherwise it would have been diluted away by the rapid spacetime expansion. In order to produce a significant asymmetry during inflation, the production rate of baryonic charge must exceed its dilution rate. Despite this challenge, it has been found that inflationary dynamics may be able to support such a scenario. By utilising the observation that, if a large baryonic charge density is created due to small-scale quantum fluctuations, it will typically be stretched out over large scales due to inflation. In the last decade mechanisms have been proposed to explore this idea, but with varying success.

In this chapter, we argue that the inflationary setting can support the generation of both the matter-antimatter asymmetry and dark matter through the extension of the SM by an anomalous gauge symmetry [1, 2]. This is achieved through the addition of a general anomalous $U(1)_X$ and a dark matter fermion candidate ψ, carrying an X charge, to the SM. The associated anomaly terms source $\mathcal{CP}$ and X charge violating processes, leading to the generation of a non-zero Chern-Simons number during inflation and subsequently a non-zero baryon number density, depending on the choice of X charge. This model is also motivated by, and able to explain, the observation that the dark matter energy density of the universe is of the same order of magnitude as that observed in the baryonic sector.

3.1 The Matter-Antimatter Asymmetry

The observed matter-antimatter asymmetry is one of the major mysteries of cosmology and particle physics. The SM predicts the existence of an asymmetry but it is many orders of magnitude smaller than that observed [3]. This asymmetry, whose energy density is believed to encompass approximately all the visible matter seen today, is quantified by the baryon asymmetry parameter,

© Springer International Publishing AG, part of Springer Nature 2018
N. D. Barrie, *Cosmological Implications of Quantum Anomalies*,
Springer Theses, https://doi.org/10.1007/978-3-319-94715-0_3

$$\eta_B = \frac{n_B - n_{\bar{B}}}{s} \simeq \frac{n_B}{s} \simeq 8.5 \cdot 10^{-11} \,, \tag{3.1}$$

where n_B ($n_{\bar{B}}$) is the baryon (antibaryon) number density and s is the entropy density of the universe. The observed value of the baryon-to-entropy ratio of our Universe is determined from the observations of the temperature fluctuations in the CMB [4–6] and from BBN predictions of the light element abundances [7–11].

The possibility of this asymmetry being an initial condition of the universe has been explored in the past, but has been found to be inconsistent with a successful inflationary epoch, other than in highly unnatural circumstances [12]. This is because the required initial energy density of the baryon asymmetry may dominate over the inflaton energy density, and hence inflation could not begin. This makes it very difficult to construct a model in which one can have an initial baryon number big enough to survive the dilution caused by inflation, and hence a dynamic generating mechanism during or after inflation is required. The compelling nature of this mystery has led to the proposal of many models for Baryogenesis; for some reviews see Refs. [13–22]. A possible connection between the inflationary setting and the dynamical generation of the observed baryon asymmetry has been considered in the past, but due to the large dilution associated with inflation it is difficult to accumulate a large enough asymmetry. Due to this, most mechanisms of Baryogenesis are assumed to have occurred after inflation; during the reheating epoch [23] or the radiation epoch, prior to BBN. In this chapter, we will consider an unorthodox mechanism which acts during inflation, utilising an anomalous gauge symmetry extension to the SM.

In order to explain the observed baryon asymmetry of the universe, any proposed mechanism must satisfy a set of criteria known as the Sakharov conditions [24].

3.1.1 The Sakharov Conditions

The Sakharov conditions [24], formulated by A.D. Sakharov in 1967, are the requirements for successful Baryogenesis in the early universe, assuming $\mathcal{CPT}$ is conserved. They are as follows:

- B **violation** If immediately after the Big Bang there was zero net baryon number and B is strictly conserved, then the net baryon number density of the universe would remain zero. Furthermore, if B is strictly conserved and there was a large initial net B this would be almost completely diluted by the end of the inflationary epoch. This is in conflict with observational evidence, hence requiring the existence of B violating interactions or mechanisms [25]. Examples of possible B violating processes are perturbative proton decay [26, 27], and non-perturbative sphaleron transitions [15].
- $\mathcal{C}$ **and** $\mathcal{CP}$ **violation** The above baryon number violating processes are also required to violate the $\mathcal{C}$ and $\mathcal{CP}$ symmetries. Consider that there are interactions that do violate baryon number e.g. $X \rightarrow a + b$, with the antimatter equivalent

$\bar{X} \rightarrow \bar{a} + \bar{b}$. If there is no C and CP violation then $\Gamma(X \rightarrow a + b) = \Gamma(\bar{X} \rightarrow \bar{a} + \bar{b})$, where Γ is the decay rate, and similarly for the reverse reactions. This means that the matter and antimatter decays will add and subtract baryon number at the same rate, leading to no net change. Therefore, C and CP violation is required for a net baryon number to be generated.

- **A period of non-equilibrium** The processes that violate B, C and CP must occur in a period of non-equilibrium. In thermal equilibrium, the corresponding forward and reverse reactions occur at the same rate, for both matter and antimatter processes. This means that even if C and CP are violated, the B violating reverse and forward reaction rates of each of the matter and antimatter processes will cancel out. The expected production of B in thermal equilibrium, at a temperature $T = \frac{1}{\beta}$, is given by,

$$
\begin{aligned}
\langle B \rangle_T = Tr(e^{-\beta H} B) &= Tr((CPT)(CPT)^{-1}e^{-\beta H} B) \\
&= Tr(e^{-\beta H}(CPT)^{-1}B(CPT)) \\
&= -Tr(e^{-\beta H} B) = -\langle B \rangle_T \ , \qquad (3.2)
\end{aligned}
$$

where CPT is the composition of the three discrete transformations C, P, and T [15]. The CPT operator commutes with H and anti-commutes with B. This implies that in thermal equilibrium the average baryon number production is zero. Hence a period of non-equilibrium is a requirement for a net baryon number to be produced. If the forward reactions were to become favoured, and the processes satisfy $\Gamma(X \rightarrow a + b) \neq \Gamma(\bar{X} \rightarrow \bar{a} + \bar{b})$, meaning C and CP violation, the baryon number each reaction produces will not cancel, with their reverse reactions or each other. Therefore, the inequality of the reaction rates for the forward processes will lead to an abundance of baryons over antibaryons, or vice-versa, in non-equilibrium conditions.

In fact, these criteria apply to not only the generation of a baryon asymmetry, but also any other particle asymmetry present in the universe. This is of interest with respect to dark matter, which may constitute an asymmetry produced in the dark matter sector. Such a possibility has interesting cosmological and phenomenological implications, and is an active area of research.

3.1.2 Asymmetric Dark Matter

The idea of a common origin of luminous and dark matter traces back to the 90s [28–30], but has received renewed interest in recent years; see a review in [31] and references therein. The major motivation to this hypothesis comes from the observation that the present-day mass density of dark matter is of the same order of magnitude as the density of visible matter [5],

$$\rho_{\rm DM} \simeq 5.5 \rho_B \ . \tag{3.3}$$

The similarity in these observed densities is perhaps an indication of a strong connection between the physics and cosmological evolution of visible and dark matter. Hence, within this picture an asymmetry of similar size may be expected to be generated among dark matter particles and antiparticles. In the model considered in this chapter, visible and dark matter are connected by a common anomalous gauged $U(1)_X$, which we introduce in addition to the gauge group of the SM. It is proposed that this anomalous gauge symmetry is responsible for the observed particle asymmetries, with the ratio of dark to luminous mass density related to how the X charge asymmetry is distributed between these two sectors.

3.2 Topological Vacuum States and the Chern-Simons Number

A concept that will be important in the cogenesis scenario, we consider in this chapter, is the Chern-Simons (CS) number. The CS number is an integer related to the winding number, which quantifies the topological non-triviality of a given vacuum state. The topological nature of the vacuum state arises due to non-trivial boundary conditions in the associated gauge theory. Full derivatives normally integrate to zero in the action integral because of the assumed boundary conditions. Although, due to the internal structure of the gauge theories involved this may not always be the case. These gauge field configurations can lead to non-trivial topology at the boundaries which are dependent on the choice of gauge, analogous to the boundary conditions in a kink solution. The vacua are each denoted with winding numbers that define certain homotopy classes, or gauge classes. It is not possible to smoothly deform an element of one homotopy class into an element of another class due to topological obstruction [32–34].

These topologically distinct vacua are degenerate, and are separated by potential barriers, analogous to a set of degenerate potential wells. Each are denoted by a CS number, which is proportional to the winding number, as depicted in Fig. 3.1. As these vacuum states are related to the boundary conditions of the full derivative terms, they do not contribute to perturbation theory, but they can provide interesting non-perturbative effects through transitions between them, which we shall discuss below.

The CS number of a given gauge field configuration, in terms of the associated gauge field A_μ^a, is defined as [35],

$$n_{CS} := n_g \frac{g^2}{32\pi^2} \int d^3x \, \epsilon^{ijk} Tr(A_i \partial_j A_k + \frac{2ig}{3} A_i A_j A_k) \ , \tag{3.4}$$

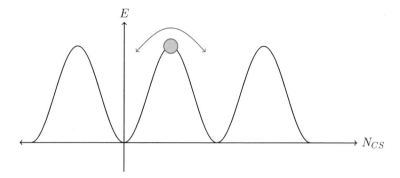

Fig. 3.1 A sphaleron transition between degenerate vacua with different CS numbers

where $A_\mu = A_\mu^a T^a$, with T^a the generators of the applicable gauge group, and g is the corresponding coupling constant.

Such topologically non-equivalent vacua are of particular relevance to the $SU(2)$ gauge group in the SM, where at infinity there is a set of mappings to the number of windings around the S^3 symmetry of the internal $SU(2)$ space. The number of windings is related to the CS number. In the SM, the $SU(2)$ theory has anomalies associated with the B and L global symmetries. This topological structure is related to these anomalies, and as we shall see, transitions between the vacua violate the associated global charges.

3.2.1 Instanton and Sphaleron Transistions

An instanton is a semi-classical solution of the gauge field equations of motion in Euclidean space, that describes a topologically non-trivial vacuum gauge field configuration [36–43]. The instanton transitions are non-perturbative processes, meaning that they cannot be described in the Lagrangian by a renormalisable operator, and as such the interaction cannot be represented by a Feynman diagram. However, an effective operator, or vertex, can be constructed. In the SM, instanton solutions are present for the weak and strong gauge groups ($SU(2)$ and $SU(3)$), and have anomalous global currents through which such transitions can be mediated.

In Electroweak Baryogenesis scenarios, the source of B violation comes from so called sphaleron transitions. These are vacuum to vacuum transitions induced by thermal excitations that result in a change in the CS number, which is depicted in Fig. 3.1. The $SU(2)$ gauge field vacuum configurations lead to the manifestation of these topological vacua, and as such the transitions between vacuum states are mediated via the B and L global anomalous currents. Each transition event results in $\Delta B = \Delta L = n_g$, meaning that $B - L$ is conserved; n_g is the number of generations. An example of such a process is given in Fig. 3.2.

Fig. 3.2 An example of a
sphaleron transition, three
antileptons are converted
into nine quarks, conserving
$B - L$

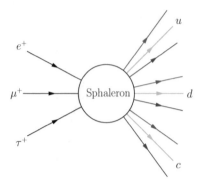

As described above, instanton quantum tunnelling processes or thermal excitations can change the topological number of a vacuum state. During such a transition, the fermion fields must also evolve as the gauge field configuration changes between topologically distinct vacuum states. This leads to the generation of particles depending on the associated anomalous symmetries.

Seeing as sphalerons correspond to thermal excitations over the potential barriers, rather than quantum tunnelling, they are highly suppressed at low temperatures, and are only important at temperatures above the characteristic energy of the potential barrier. In the case of the electroweak vacuum, the sphaleron processes become important at temperatures above the EWPT ($T \sim 100$ GeV). The thermal rate of sphaleron processes in the broken phase, below the temperature of the EWPT, is [44],

$$\Gamma_{sph}(T) = \mu \left(\frac{M_W}{\alpha_W T} \right)^3 M_W^4 e^{\frac{E_{sph}(T)}{T}} , \tag{3.5}$$

where M_W is the mass of the W boson, T is the temperature, μ is a constant, $E_{sph}(T)$ is the sphaleron energy, and $\alpha_W = \frac{g_2^2}{4\pi} \simeq \frac{1}{29}$ with g_2 being the gauge coupling of $SU(2)$. This illustrates why these baryon and lepton number violating processes are not observed today. An estimate of the rate of sphaleron transitions per unit volume, in which κ is a dimensionless constant, is given by,

$$\Gamma_{sph}(T) \simeq \kappa \alpha_W^4 T^4 . \tag{3.6}$$

This rate is strongly dependent on the temperature of the state, and are in thermal equilibrium when the temperatures are between the EWPT ($T \sim 100$ GeV) and 10^{12} GeV the sphalerons [45, 46]. Therefore, these processes are of interest when investigating early universe particle dynamics, due to the high temperatures that may be present.

Although not integral to our scenario, electroweak sphaleron transitions need to be taken into account. They can act in thermal equilibrium after reheating, leading to B and L redistribution. Depending on the reheating temperature, the $B - L$ number contained within the X charge asymmetry generated by our mechanism will be redistributed between the SM fermionic degrees of freedom by equilibrium sphaleron transitions during the radiation era. The resultant baryon number density distributed from the initial $B - L$ number density is given by the following [47],

$$n_B = \frac{28}{79} n_{B-L} , \qquad (3.7)$$

where this relation is applicable when the temperature of the primordial plasma is above $T \sim 100$ GeV.

The time variation of the vacuum state in an inflationary setting can lead to the accumulation of CS number in the presence of the X charge anomalies. In the scenario we propose in this work, the accumulated topological charge in the vacuum will thus correspond to a net X charge.

3.3 A Model of Inflationary Cogenesis

In what follows, we consider an application of these ideas in an inflationary setting to construct a new mechanism for generating both luminous and dark matter during cosmic inflation. It has been suggested that inflation may play an even more prominent role in cosmology than solving the mysteries discussed in Chap. 1, but by also generating the observed matter-antimatter asymmetry in the universe [1, 48–52]. Namely, in [1] we argued that a successful Baryogenesis scenario can be realised during inflation within models containing anomalous gauge symmetries [53]. One of the first uses of anomalies for the generation of the baryon asymmetry is in a model of inflationary leptogenesis [49], in which a lepton asymmetry is produced during inflation due to the gravitational birefringence induced through the gravitational lepton number anomaly coupled to a new pseudoscalar field.

In the model we consider in this chapter, ordinary and dark matter both carry charges associated with an anomalous $U(1)_X$ group. Anomaly terms in the model Lagrangian source $\mathcal{CP}$ and $U(1)_X$ charge violating processes during inflation, producing corresponding non-zero CS numbers which are subsequently reprocessed into baryon and dark matter densities. Other recent works have considered relating the generation of the luminous matter-antimatter asymmetry with an asymmetry in the dark sector within a gauged $U(1)_X$ extension of the SM [54–60]. In the early universe, when the expansion rate is faster than processes with fermion chirality flip, the gauged anomalies may effectively appear within the SM [61]. Indeed, it has been argued that anomalous production of the right-handed electron number is possible through the hypercharge anomaly in the SM [62]. An inflationary version of the above scenario is discussed in [50]. The anomalous gauge theory we consider can also be viewed as an effective low-energy theory, which admits a fundamental completion free of gauge anomalies. The obvious candidates for such an anomalous gauge theory are gauged baryon and lepton numbers, or any linear combination thereof.

As with any $\mathcal{CPT}$ invariant model attempting to explain the dynamical generation of a charge asymmetry, it must satisfy the Sakharov's conditions as discussed
above, and ours does so as follows. The anomalies present upon introduction of
the $U(1)_X$ gauge boson provide our X charge violation. On top of this, the gauge
invariance of the $U(1)_X$ requires a pseudoscalar field to couple to these anomalies,
which describes the longitudinal polarization of the X boson. In the cosmological
setting these interactions spontaneously violate $\mathcal{CP}$ invariance, and the inflationary
epoch provides the push out-of-equilibrium that is required for the accumulation of
the asymmetry. This will be discussed in more detail below. Note that this mechanism differs from the one presented in [63], in which a non-zero CS number in the
hypercharge field is generated during inflation. In their case, the conversion of this to
baryon number happens at the electroweak scale through altering the right-handed
electron chemical potential, while also requiring a strongly first order electroweak
phase transition which is not achieved within the SM.

3.3.1 Models with an Anomalous $U(1)_X$

In this general mechanism, we consider an extension of the SM that is based
on the $SU(3) \times SU(2) \times U(1)_Y \times U(1)_X$ gauge group and contains an additional
fermion(s) that shall act as a dark matter candidate. The introduction of a scalar to
play the role of the inflaton is also required, but the detailed dynamics of the inflationary epoch is not important for our analysis. The new $U(1)_X$ gauge symmetry is
assumed to be anomalous, and hence the corresponding gauge boson will be necessarily massive with gauge invariance realised non-linearly. The longitudinal degree
of freedom of this $U(1)_X$ gauge field is then described by a scalar field $\theta(x)$, which
allows anomaly cancellation through the introduction of appropriate counter-terms
[53], through the Green-Schwarz mechanism as discussed in Chap. 1. This theory
can be viewed as a low energy limit of an anomaly-free theory, either within ordinary QFT or string theory. In the presence of a cubic anomaly $U(1)_X^3$, the additional
Lagrangian terms important to our analysis are,

$$\frac{1}{\sqrt{-g}}\mathcal{L}_X = -\frac{1}{4}g^{\mu\alpha}g^{\nu\beta}X_{\mu\nu}X_{\alpha\beta} + \frac{1}{2}f_X^2 g^{\mu\nu}\left(g_X X_\mu - \partial_\mu\theta\right)\left(g_X X_\nu - \partial_\nu\theta\right)$$

$$- \mathcal{A}_1 \frac{g_X^2}{16\pi^2}\theta(x)X_{\mu\nu}\tilde{X}^{\mu\nu}\,, \tag{3.8}$$

where $X_{\mu\nu}$ denotes the field strength of the $U(1)_X$ gauge boson with corresponding
coupling constant $g_X = m_X/f_X$, f_X is a parameter that defines the mass of the $U(1)_X$
boson (m_X), and $\tilde{X}^{\mu\nu} = \frac{1}{2\sqrt{-g}}\epsilon^{\mu\nu\rho\sigma}X_{\rho\sigma}$ is the dual field strength, in which $\epsilon^{\mu\nu\rho\sigma}$ is
the Levi-Civita tensor. We have omitted fermion interactions and the charged current
j_X terms. The final term in Eq. (3.8) is responsible for maintaining gauge invariance
of the full quantum theory description under $U(1)_X$ transformations, as discussed

Table 3.1 The representations of the SM fermions and dark fermion ψ in reference to the gauge symmetries of the theory

Fermion field	$SU(3)$	$SU(2)$	$U(1)_Y$	Case 1: $U(1)_{B-L}$	Case 2: $U(1)_B$
$Q_L^i = \begin{pmatrix} u \\ d \end{pmatrix}_L^i$	3	2	1/6	1/3	1/3
u_R^i	3	1	2/3	1/3	1/3
d_R^i	3	1	$-1/3$	1/3	1/3
$L^i = \begin{pmatrix} \nu \\ e \end{pmatrix}_L^i$	1	2	$-1/2$	-1	0
e_R^i	1	1	-1	-1	0
ψ	1	1	0	q_ψ	q_ψ

above. In the proceeding analysis any associated gravitational anomaly is ignored as its contribution is considered to be negligible with respect to the other anomalous contributions.

We shall consider two example applications of this general model for cogenesis. Namely, we consider the gauged $B - L$ and B extensions both of which are anomalous [64–74]. The examples we discuss shall also contain a fermionic field(s) ψ, which carries a chiral charge under the anomalous gauge symmetry and is sterile under the SM gauge symmetry. The charges of each of the fermions under these additional gauge symmetries are given in Table 3.1. The mass m_ψ is an extra parameter which can be directly introduced within the non-linear realisation of the anomalous gauge symmetry. Typically, such a mass is also generated radiatively within the low-energy effective theory, reflecting a more conventional mechanism for mass generation within an ultraviolet anomaly-free completion.

The stability of the dark matter candidate ψ is ensured as the Lagrangian in Eq. (3.8) has no interaction vertices which allow violation of the global ψ number locally. Violation of X charge, and hence also the dark X charge, only occurs in this model through the non-perturbative generation of CS number in an expanding spacetime. This production is only of significance during inflation, as contributions during subsequent matter and radiation dominated epochs are negligible due to minimal $\mathcal{CP}$ violation and push out-of-equilibrium. This means that after the inflationary epoch ends, the dark matter density to entropy density ratio is fixed, assuming no significant additional sources of entropy.

Case 1: $U(1)_{B-L}$ **and a sterile fermion** ψ

In the SM, the $U(1)_{B-L}$ symmetry is anomalous unless three right handed neutrinos are introduced. The associated anomalies are trace and cubic: $\mathcal{A}_0(U(1)_{B-L}) = -3$ and $\mathcal{A}_1(U(1)_{B-L}^3) = -3$. We introduce N_ψ new right-handed (for definiteness) Weyl fermions ψ, some of which act as dark matter candidates in our model. For simplicity we assume that they carry the same $B - L$ charge q_ψ and interact only via exchange of the $B - L$ gauge boson. The addition of these states alters the $B - L$ anomalies as follows: $\mathcal{A}_0 := \mathcal{A}_0(U(1)_{B-L}) = -N_\psi q_\psi - 3$ and $\mathcal{A}_1 := \mathcal{A}_1(U(1)^3_{B-L}) = -N_\psi q^3_\psi - 3$. In this case, the dark matter fermion does not introduce any new anomalies. We will ignore the gravitational anomaly $\mathcal{A}_0$ in our analysis, but it should be noted that taking $q_\psi = -3$ or -1 and $N_\psi = 1$ or 3, respectively, eliminates $\mathcal{A}_0$. Obviously, $N_\psi = 3$, $q_\psi = -1$ removes all anomalies, so we are not interested in such a charge assignment in this paper.

The addition of the $B - L$ gauge symmetry and dark matter candidate to the SM leads to a Lagrangian density of the same form given for the general case presented in Eq. (3.8).

Case 2: $U(1)_B$ **and a sterile baryon** ψ
Gauging the baryon number symmetry of the SM results in the inclusion of two mixed anomalies involving the weak and hypercharge gauge groups: $\mathcal{A}_2(SU(2)^2 \times U(1)_B) = 3/2$ and $\mathcal{A}_3(U(1)^2_Y \times U(1)_B) = -3/2$. The addition of a new sterile state ψ leaves these mixed anomalies unchanged, but introduces two new unmixed anomalies: $\mathcal{A}_0 := \mathcal{A}_0(U(1)_B) = -N_\psi q_\psi$ and $\mathcal{A}_1 := \mathcal{A}_1(U(1)^3_B) = -N_\psi q^3_\psi$. Hence, there are four anomalies, each of which will contribute to baryonic charge generation during the inflationary epoch, but only two of which will include generation of fermions in the dark matter sector, namely, $\mathcal{A}_0$ and $\mathcal{A}_1$.

The presence of additional mixed anomalies means that extra anomaly cancelling terms are required with respect to the gauged $B - L$ case considered above, that is,

$$\frac{1}{\sqrt{-g}}\mathcal{L}_X = -\frac{1}{4}g^{\mu\alpha}g^{\nu\beta}X_{\mu\nu}X_{\alpha\beta} + \frac{1}{2}f^2_X g^{\mu\nu}\left(g_X X_\mu - \partial_\mu\theta\right)(g_X X_\nu - \partial_\nu\theta)$$
$$- \mathcal{A}_1\frac{g^2_X\theta(x)}{16\pi^2}X_{\mu\nu}\tilde{X}^{\mu\nu} - \mathcal{A}_2\frac{g^2_1\theta(x)}{16\pi^2}B_{\mu\nu}\tilde{B}^{\mu\nu} - \mathcal{A}_3\frac{g^2_2\theta(x)}{16\pi^2}W^a_{\mu\nu}\tilde{W}^{a\mu\nu},$$
$$(3.9)$$

where $B_{\mu\nu}$ and $W_{\mu\nu}$ denote the hypercharge and weak field strengths respectively, with corresponding coupling constants g_1 and g_2.

3.4   Dynamics of an Anomalous Gauge Field During Inflation

For a model to successfully produce a charge asymmetry in the early universe it must satisfy the well-known Sakharov conditions [24]. We will now discuss the framework of our new mechanism for cogenesis and how it satisfies these criteria in more detail.

Firstly, we wish to describe the universe using the Friedmann-Robertson-Walker metric tensor, which represents a homogeneous, isotropic and spatially flat cosmological spacetime. In conformal coordinates the metric can be expressed as: $g_{\mu\nu} = a^2(\tau)\eta_{\mu\nu}$. During inflation the scale factor $a(\tau)$ is given by the following,

$$a(\tau) = -1/H_{\text{inf}}\tau,\qquad\qquad(3.10)$$

where H_{inf} is the expansion rate during inflation ($H_{\text{inf}} \cong$ constant). The conformal time is given in the range $\tau \in [-\frac{1}{H}, -\frac{1}{H}e^{-N_{\text{inf}}}]$ during inflation, where N_{inf} is the number of e-folds during the inflationary epoch, such that the scale factor is $a(\tau_0) = 1$ at the beginning of inflation.

To allow analytical treatment, the analysis that follows requires certain simplifying assumptions. For the θ field we only consider a classical homogeneous background configuration, $\theta(\tau, \vec{x}) = \theta(\tau)$, and ignore quantum fluctuations over it. We take $g_X \ll 1$ such that the $\theta(x)$ and $X_\mu(x)$ fields essentially decouple from each other. This also implies that the $U(1)_X$ boson is light relative to the scale f_X, $m_X/f_X \ll 1$, and hence we will not be interested in its dynamics during inflation. With these assumptions the Lagrangian in Eq. (3.8) becomes,

$$\mathcal{L}_X = -\frac{1}{4}\eta^{\mu\alpha}\eta^{\nu\beta}X_{\mu\nu}X_{\alpha\beta} + \frac{1}{2}a(\tau)^2\eta^{\mu\nu}\left(m_X X_\mu - \partial_\mu\phi(\tau)\right)\left(m_X X_\nu - \partial_\nu\phi(\tau)\right)$$
$$- \mathcal{A}_1 \frac{g_X^2\phi(\tau)}{32\pi^2 f_X}\epsilon^{\mu\nu\alpha\beta}X_{\mu\nu}X_{\alpha\beta}, \tag{3.11}$$

where $\phi(\tau) \equiv f_X\theta(\tau)$. From this Lagrangian follows the equation of motion for $\phi(\tau)$,

$$\left(a^2\phi'\right)' = 0, \tag{3.12}$$

where $\phi' \equiv d\phi/d\tau$ and we have ignored any terms quadratic in X_μ. Solving for $\phi'(\tau)$ we obtain,

$$\phi'(\tau) = \frac{a^2(\tau_0)\phi_0'}{a^2(\tau)}, \tag{3.13}$$

where ϕ_0' is an integration constant associated with the 'field velocity' at the start of inflation, which is defined at $\tau = \tau_0$, where $a(\tau_0) = 1$. An upper limit on the value of ϕ_0' is provided by the requirement that the initial energy density of the ϕ field be less than that of the inflaton field. This upper limit is thus $\phi_0' \lesssim H_{\text{inf}} M_p$, where $M_p = 1/\sqrt{G}$ is the Planck mass. Substituting Eq. (3.13) into the linearised equation of motion for the X_μ gauge field gives,

$$\left(\partial_\tau^2 - \vec{\nabla}^2 + \left(\frac{a(\tau_0)m_X}{H_{\text{inf}}\tau}\right)^2\right)X^i + \kappa_X\tau^2\epsilon^{ijk}\partial_j X_k = 0, \tag{3.14}$$

where

$$\kappa_X = |\mathcal{A}_1|\frac{a^2(\tau_0)g_X^2\phi_0'H_{\text{inf}}^2}{4\pi^2 f_X}, \tag{3.15}$$

and the gauge $X_0 = \partial_i X_i = 0$ has been chosen. The source of $\mathcal{CP}$ violation in our model is apparent in Eq. (3.14) where the two terms have opposite $\mathcal{P}$, and hence, $\mathcal{CP}$ transformations.

In the discussion that follows we treat the $U(1)_X$ gauge boson as a massless particle, as we have assumed $m_X \ll H_{\text{inf}}$. To then quantize this model we promote the X gauge boson fields to operators and assume that the boson has two possible circular polarisation states,

$$X_i = \int \frac{d^3\vec{k}}{(2\pi)^{3/2}} \sum_\alpha \left[G_\alpha(\tau, k)\epsilon_{i\alpha}\hat{a}_\alpha e^{i\vec{k}\cdot\vec{x}} + G_\alpha^*(\tau, k)\epsilon_{i\alpha}^*\hat{a}_\alpha^\dagger e^{-i\vec{k}\cdot\vec{x}} \right] , \qquad (3.16)$$

where $\vec{\epsilon}_\pm$ denotes the two possible helicity states of the $U(1)_X$ gauge boson ($\vec{\epsilon}_+^* = \vec{\epsilon}_-$) and the creation, $\hat{a}_\alpha^\dagger(\vec{k})$, and annihilation, $\hat{a}_\alpha(\vec{k})$, operators satisfy the canonical commutation relations,

$$\left[\hat{a}_\alpha(\vec{k}), \hat{a}_\beta^\dagger(\vec{k}') \right] = \delta_{\alpha\beta}\delta^3(\vec{k} - \vec{k}') , \qquad (3.17)$$

and

$$\hat{a}_\alpha^a(\vec{k})|0\rangle_\tau = 0 , \qquad (3.18)$$

where $|0\rangle_\tau$ is an instantaneous vacuum state at time τ.

The mode functions in Eq. (3.16) are described by the following equations, from Eq. (3.14),

$$G_\pm'' + \left(k^2 + \frac{\lambda^2}{\tau^2} \mp \kappa_X \tau^2 k \right) G_\pm = 0 , \qquad (3.19)$$

where $\lambda = \frac{m_X}{H_{\text{inf}}}$, which is assumed to be small as stated above.

Solving for the mode functions $G_\pm$ in Eq. (3.19) gives,

$$G_+(\tau, k) = 2^{\frac{1+\nu}{2}} e^{-z} 2\tau^{\frac{1}{2}+\nu} \left[C_1 U\left(\frac{1+\nu}{2} - \frac{\Omega_k}{4}, 1+\nu, z \right) + C_2 M\left(\frac{1+\nu}{2} - \frac{\Omega_k}{4}, 1+\nu, z \right) \right] \qquad (3.20)$$

and

$$G_-(\tau, k) = 2^{\frac{1+\nu}{2}} e^{z} \tau^{\frac{1}{2}+\nu} \left[C_3 U\left(\frac{1+\nu}{2} - \frac{i\Omega_k}{4}, 1+\nu, \frac{z}{i} \right) + C_4 M\left(\frac{1+\nu}{2} - \frac{i\Omega_k}{4}, 1+\nu, \frac{z}{i} \right) \right] \qquad (3.21)$$

where $z = \frac{k^2\tau^2}{\Omega_k}$, $\Omega_k = \sqrt{\frac{k^3}{\kappa_X}}$, $\nu = \frac{1}{2}\sqrt{1 - 4\lambda^2} \sim \frac{1}{2} - \lambda^2$, $U(a, b, z)$ is a confluent hypergeometric function of the second kind, and $M(a, b, z)$ is a confluent hypergeometric function of the first kind (Kummer Function).

In the limit $|\tau| \to 0$ (or $k^2 + \frac{\lambda^2}{\tau^2} \gg \kappa_X \tau^2 k$), $\mathcal{CP}$-invariant wave modes are obtained. These are described by,

$$X_i = \int \frac{d^3\vec{k}}{(2\pi)^{3/2}} \sum_\alpha \left[F_\alpha(\tau, k)\epsilon_{i\alpha}\hat{b}_\alpha e^{i\vec{k}\cdot\vec{x}} + F_\alpha^*(\tau, k)\epsilon_{i\alpha}^*\hat{b}_\alpha^\dagger e^{-i\vec{k}\cdot\vec{x}} \right] , \qquad (3.22)$$

where the wave mode functions $F_\pm$ are found to be,

$$F_+(\tau, k) = \frac{\sqrt{\pi \tau}}{2} H_\nu^{(2)}(k\tau) e^{-i\frac{\pi}{2}(\frac{1}{2}+\nu)} \quad \text{and} \quad F_-(\tau, k) = \frac{\sqrt{\pi \tau}}{2} H_\nu^{(1)}(k\tau) e^{i\frac{\pi}{2}(\frac{1}{2}+\nu)} .$$

$$(3.23)$$

By matching the modes in Eqs. (3.20) and (3.21) to those in Eq. (3.23) and using the known Wronskian normalisation we can determine the coefficients C_{1-4}. For more details on this calculation and the form of the coefficients see Appendix B.

Particle Creation during Inflation and Bogolyubov Transformations

In an expanding spacetime it is difficult to define a time independent vacuum state because the Hamiltonian becomes time dependent. In a flat spacetime, the vacuum is defined with reference to plane wave solutions, such that excitations from the vacuum state correspond to plane waves. In an expanding spacetime background the form of the annihilation and creation operators will change with time, and thus the canonical vacuum state can only be defined at any given instant in time. Thus, the Hamiltonian's instantaneous energy eigenstates will not be the same at all times, and hence the vacuum state will not be as well. That is, the vacuum state at one time will correspond to an excited state of the vacuum at another time, which means that on comparing the vacua, a relative particle number with reference to one another may be found. This is how the expansion of the universe during inflation can lead to particle production [75–78]. It is possible to relate the particle number of two vacuum states by a so called Bogolyubov transformation. This type of transformation allows the determination of the particle content of an evolving vacuum state with respect to another vacuum state.

If we define two different annihilation operators $\hat{a}$ and $\hat{b}$ we in turn define the corresponding a and b vacua; $|0_a\rangle$ and $|0_b\rangle$, respectively. It is possible to express the b-vacuum as a superposition of a particle states and vice versa. The a-vacuum will contain no a particles, but may have a b-particle density. A Bogolyubov transformation performed between the a-vacuum and b-vacuum can allow the calculation of this relative particle density, and is defined as [79],

$$\hat{b}_\gamma = \alpha_\gamma \hat{a}_\gamma + \beta_\gamma^* \hat{a}_\gamma^\dagger \quad \text{and} \quad \hat{b}_\gamma^\dagger = \alpha_\gamma^* \hat{a}_\gamma^\dagger + \beta_\gamma \hat{a}_\gamma , \tag{3.24}$$

where α and β are k dependent complex numbers. In order to determine α and β the modes corresponding to each vacuum state must be matched at some point in their evolution. This is done as follows,

$$v^*(t_0) = \alpha_\gamma u_\gamma^*(t_0) + \beta_\gamma u_\gamma(t_0) , \tag{3.25}$$

$$v^{*'}(t_0) = \alpha_\gamma u_\gamma^{*'}(t_0) + \beta_\gamma u_\gamma'(t_0) , \tag{3.26}$$

where v and u are the modes defined for the a and b vacuum states, respectively.

In our model, a Bogolyubov transformation will be utilised to match the evolving inflationary vacuum and the vacuum state for super horizon modes. This is because, at scales smaller than the Hubble rate the effects of spacetime curvature become negligible, and solutions must converge to plane waves. Rather than matching the vacua to calculate the net particle density accumulated, we shall determine the accumulated

CS number density that is induced by the X charge violating anomalous interactions during the inflationary epoch.

In our scenario, we compare the birefringent and $\mathcal{CP}$-invariant modes to derive the Bogolyubov coefficients relating the two sets of creation and annihilation operators, $\{\hat{a}^a_\alpha, \hat{a}^{a\dagger}_\alpha\}$ and $\{\hat{b}^a_\alpha, \hat{b}^{a\dagger}_\alpha\}$, in Eqs. (3.16) and (3.22). The Bogolyubov transformations in this case are defined by,

$$\hat{b}^a_\alpha(\vec{k}) = \alpha_\alpha a^{a\dagger}_\alpha(\vec{k}) + \beta^*_\alpha \hat{a}^a_\alpha(\vec{k}) , \tag{3.27}$$

$$\hat{b}^{a\dagger}_\alpha(\vec{k}) = \alpha^*_\alpha a^a_\alpha(\vec{k}) + \beta_\alpha \hat{a}^{a\dagger}_\alpha(\vec{k}) . \tag{3.28}$$

The relevant Bogolyubov coefficients are found to be,

$$\alpha_\pm = 1 - \frac{1}{2^{1-\nu}} \left(1 \pm \frac{i\lambda^2}{(k\tau)^{1-2\lambda^2}} \left(1 - \frac{\pi(k\tau)^{1-2\lambda^2}}{2^\nu} \right) \mp \frac{i2^{1-\nu}(k\tau)^{\lambda^2}}{\sqrt{k}} e^{\mp i\pi\lambda^2/2} G'^*_\pm \big|_{\frac{k\tau^2}{k}, k|\tau|\to 0} \right) , \tag{3.29}$$

and,

$$\beta_\pm = \frac{e^{\mp i\pi\lambda^2}}{2^{1-\nu}} \left(1 \mp \frac{i\lambda^2}{(k\tau)^{1-2\lambda^2}} \left(1 - \frac{\pi(k\tau)^{1-2\lambda^2}}{2^\nu} \right) \pm \frac{i2^{1-\nu}(k\tau)^{\lambda^2}}{\sqrt{k}} e^{\pm i\pi\lambda^2/2} G'^*_\mp \big|_{\frac{k\tau^2}{k}, k|\tau|\to 0} \right) , \tag{3.30}$$

where we have considered the superhorizon modes ($k|\tau| \approx 0$).

3.5 Simultaneous Generation of Luminous and Dark Matter During Inflation

Now that we have determined the dynamics of the X_μ gauge field we can calculate the general X charge density generated during inflation. It is known that the anomalous non-conservation of the X charge current is given by,

$$\partial_\mu \left(\sqrt{-g} j^\mu_X \right) = \mathcal{A}_1 \frac{g^2_X}{32\pi^2} \epsilon^{\mu\nu\rho\sigma} X_{\mu\nu} X_{\rho\sigma} \equiv \mathcal{A}_1 \frac{g^2_X}{8\pi^2} \partial_\mu \left(\sqrt{-g} K^\mu \right) , \tag{3.31}$$

where $K^\mu = \frac{1}{2\sqrt{-g}} \epsilon^{\mu\nu\rho\sigma} X_{\nu\rho} X_\sigma$ is a topological current. This implies that the net X charge density $n_X = n_x - n_{\bar{x}} \equiv a^{-1}(\tau)\langle 0|j^0_X|0\rangle$ is related to the CS number density of the $U(1)_X$ gauge boson by the following equation,

$$n_X = |\mathcal{A}_1| \frac{g^2_X}{8\pi^2} a(\tau_{\text{end}}) n_{CS} , \tag{3.32}$$

where $\tau = \tau_{\text{end}}$ is the conformal time at the end of inflation, and $n_X(\tau_0) = n_{CS}(\tau_0) = 0$ at the start of inflation. The form of the CS number is given below, in which we wish to consider only large scale superhorizon modes ($k|\tau| \simeq 0$),

$$n_{CS} = \frac{1}{a^4(\tau_{\text{end}})} \epsilon^{ijk} \lim_{k|\tau| \to 0} \langle 0|X_i \partial_j X_k|0\rangle$$

$$\simeq \frac{1}{4\pi^2 a^4(\tau_{\text{end}})} \int_\mu^\Lambda k dk \left[|G'_+|^2_{\frac{\kappa\tau^2}{k}, k|\tau| \to 0} - |G'_-|^2_{\frac{\kappa\tau^2}{k}, k|\tau| \to 0} \right] - \mathcal{O}(\lambda^2), \tag{3.33}$$

where we ignore small terms with quadratic or higher orders of λ. Note that the upper limit in the integral in Eq. (3.33) simply cuts out sub-horizon modes for which our approximate calculations are not applicable. The ultraviolet modes do not give a significant contribution anyway, since they act as $\mathcal{CP}$-invariant planewaves, which expectantly lead to a cancellation between the positive and negative frequency modes. The dominant contribution to n_{CS} is given by infrared modes, and in fact the integral is divergent. This divergence is reminiscent of the well-known infrared divergence of de Sitter-invariant two-point functions, which possibly signals that the pure de Sitter approximation of the inflationary phase becomes inadequate in our case. There is no commonly accepted prescription for regularization of these types of divergences in the literature and we simply introduce an infrared cut-off μ.

We assume that the only non-negligible source of entropy density is the process of reheating after inflation, for which the entropy density produced is $s \simeq \frac{2\pi^2}{45} g_* T_{\text{rh}}^3$, where T_{rh} is the associated reheating temperature and $g_*(T_{\text{rh}}) \simeq 106.75$. Upon taking a first order expansion around $\Omega_k = 0$ in Eq. (3.33), we obtain the following expression for the X charge asymmetry parameter generated by the unmixed anomaly,

$$\eta_X = \frac{n_X}{s} \approx |\mathcal{A}_1| \frac{30 g_X^2}{\pi^{10} g_*} \Gamma \left(\frac{3}{4}\right)^4 e^{-3N_e} \left(\frac{\kappa_X}{\mu T_{\text{rh}}^2}\right)^{\frac{3}{2}}$$

$$\approx 8.5 \cdot 10^{-11} |\mathcal{A}_1|^{5/2} \left(\frac{m_X}{10^{12} \text{ GeV}}\right)^5 \left(\frac{\phi_0'}{10^{32} \text{ GeV}^2}\right)^{\frac{3}{2}} \left(\frac{H}{10^{14} \text{ GeV}}\right) \tag{3.34}$$

$$\times \left(\frac{T_{\text{rh}}}{2 \cdot 10^{11} \text{ GeV}}\right)^{-2} \left(\frac{f_X}{10^{14} \text{ GeV}}\right)^{-\frac{13}{2}} \left(\frac{\mu}{10^{-42} \text{ GeV}}\right)^{-\frac{3}{2}},$$

where N_e denotes the minimum number of e-folds required to solve the horizon and flatness problems, and includes the additional dilution that occurs if the reheating period is not instantaneous. The number of e-folds that contribute to the dilution of n_X is,

$$N_e = N_{\text{inf}} + N_{\text{rh}} \simeq 27.5 + \frac{2}{3} \ln \left(\frac{H_{\text{inf}} M_p}{(1 \text{ GeV})^2}\right) - \frac{1}{3} \ln \left(\frac{T_{\text{rh}}}{1 \text{ GeV}}\right), \tag{3.35}$$

where

$$N_{\text{inf}} \simeq 34 + \ln \left(\frac{T_{\text{rh}}}{100 \text{ GeV}}\right) \quad \text{and} \quad N_{\text{rh}} \simeq \frac{2}{3} \ln \left(\frac{H_{\text{inf}} M_p}{T_{\text{rh}}^2}\right) - 1.89. \tag{3.36}$$

In Eq. (3.34), it can be seen that taking the parameter κ_X to be large increases the asymmetry. Also, the infrared cut-off μ must be sufficiently small. In what follows we consider two possible cut-offs - the minimal box cut-off [80], $\mu = H_0 \approx 10^{-42}$ GeV, which accounts for all the modes that are within the present Hubble horizon, and $\mu = H_{\mathrm{BBN}} \approx 10^{-25}$ GeV which includes all of the modes within the Hubble horizon at the beginning of the period of BBN; assuming $T_{\mathrm{BBN}} \sim 1$ MeV.

A similar relation to Eq. (3.34) can be derived for the mixed anomalies. In particular, these can be present in the case of gauged baryon number $U(1)_B$, as considered in [1]. In the case of the electroweak and hypercharge mixed anomalies the extra contribution to the total X charge asymmetry is,

$$
\eta_X^{\mathrm{mixed}} = \frac{n_X^{\mathrm{mixed}}}{s} \approx (|\mathcal{A}_2|^{5/2} g_1^5 + 3|\mathcal{A}_3|^{5/2} g_2^5) \frac{15}{4\pi^{13} g_*} \Gamma\left(\frac{3}{4}\right)^4 e^{-3N_e} \left(\frac{\kappa}{\mu T_{\mathrm{rh}}^2}\right)^{\frac{3}{2}},
$$
(3.37)

where $\kappa = \frac{\phi_0' H_{\mathrm{inf}}^2}{f_X}$, and we have assumed that they have the same IR cut-off μ.

In the derivation of the asymmetry parameters above, Eqs. (3.34) and (3.37), we have assumed that the only non-negligible contribution to the generated charge asymmetry is produced during the inflationary epoch. The conditions for the mechanism considered here may still be active during the radiation epoch, but the overall effect will be negligible as the push out-of-equilibrium is considered to be too small in later epochs; hence the total X charge is assumed to be conserved once inflation ends. One exception to this is the possibility of sphaleron redistribution which will violate both the SM B and L charges equally. The mutual dilution of the charge and the entropy densities, after reheating, ensures there is no further dilution of the asymmetry parameter. No additional washout processes have been considered in the above derivation.

In the following section we utilise the known properties of sphaleron transitions to determine the distribution of X charges amongst fermionic species after the EWPT, if the reheating temperature is greater than the critical temperature ($T_c \sim 100$ GeV). How the sphaleron processes redistribute the X charge is dependent on the specific model being considered—the type of charge gauged, the associated anomalies, and the properties of the new fermion(s) introduced.

3.6   Replicating the Observed $\rho_{\mathrm{DM}}/\rho_B$ and η_B

Now that the X charge asymmetry parameter has been calculated we can determine under what conditions this model will replicate simultaneously the observed dark to luminous matter mass density ratio and the baryon asymmetry parameter. The generated X charge density can be decomposed into SM and dark matter components as follows,

$$
n_X = n_X^{SM} + n_D,
$$
(3.38)

The SM component will have an associated $B - L$ charge which will be reprocessed by the action of sphaleron transitions, before or at the EWPT, into a known fermionic distribution. The dark matter candidate considered here will be unaffected by the sphaleron transitions as it is assumed here to be a singlet under the electroweak interactions, although this does not have to be the case.

After the EWPT the $B - L$ charge will be distributed between B and L charges as follows; $(B - L)_{SM} = \frac{79}{28}B$ and $(B - L)_{SM} = -\frac{79}{51}L$. We require that the resultant SM baryon number asymmetry is consistent with that which is observed, given in Eq. (3.1). The baryon asymmetry parameter will be given by the following relation,

$$\eta_B = \epsilon(\eta_X^{\rm mixed-SM} + \eta_X^{\rm unmixed-SM}) , \tag{3.39}$$

where ϵ is a step function defined by,

$$\epsilon := \epsilon(T_{\rm rh}) = \begin{cases} \frac{28}{79} & T_{\rm rh} > T_{\rm c} \\ 1 & T_{\rm rh} < T_{\rm c} \end{cases} , \tag{3.40}$$

Henceforth we will assume that the mixed anomalies only contribute to the SM sector, $\eta_X^{\rm mixed} = \eta_X^{\rm mixed-SM}$, as our dark matter candidate is sterile under the SM gauge groups.

It is assumed that the X charge density generated is initially uniformly distributed between each of the applicable fermion degrees of freedom, that is,

$$\eta_X^{SM} = \eta_X^{\rm mixed} + \frac{\sum_i N_{SM}^i |q_{SM}^i|}{\sum_i N_{SM}^i |q_{SM}^i| + \sum_i N_D^i |q_D^i|} \eta_X^{\rm unmixed} , \tag{3.41}$$

$$\eta_D = \frac{\sum_i N_D^i |q_D^i|}{\sum_i N_{SM}^i |q_{SM}^i| + \sum_i N_D^i |q_D^i|} \eta_X^{\rm unmixed} , \tag{3.42}$$

where the index i corresponds to the particle species, N_i is the corresponding number of degrees of freedom, and q_i is the associated X charge. Therefore, the baryon asymmetry parameter defined above is given by,

$$\eta_B = \epsilon \eta_X^{SM} = \epsilon \left(\eta_X^{\rm mixed} + \frac{\sum_i N_{SM}^i |q_{SM}^i|}{\sum_i N_{SM}^i |q_{SM}^i| + \sum_i N_D^i |q_D^i|} \eta_X^{\rm unmixed} \right) . \tag{3.43}$$

The dark matter to luminous matter mass density ratio is given by,

$$\frac{\rho_D}{\rho_B} = \frac{m_\psi \, q_B \eta_D}{m_B \, q_\psi \eta_B} = \frac{\eta_D}{q_\psi \eta_B} \left(\frac{m_\psi}{1 \, \text{GeV}} \right) , \tag{3.44}$$

where we have assumed $m_B = 1$ GeV and $q_B = 1$. In the following analysis we will assume that this ratio is fixed to the observed value given in Eq. (3.3). Hence upon considering parameters that give the correct η_B we will also obtain the observed relic dark matter abundance. Now we wish to consider this framework in the two scenarios introduced earlier; namely, a gauged $B - L$ and a gauged B number, each including a single sterile fermion charged under the given group.

Case 1: $U(1)_{B-L}$ **and a sterile fermion**

In this scenario we must sum over all the SM fermions, $\sum_i N_{SM}^i |q_{SM}^i| = 21$, assuming no RH neutrinos have been added. Only the unmixed cubic anomaly contributes to the $B - L$ charge generation. Using these facts and Eq. (3.44) we derive the following dark matter to luminous matter mass density ratio,

$$\frac{\rho_D}{\rho_B} \approx \frac{1}{21\epsilon} \left(\frac{m_\psi}{1 \text{ GeV}} \right) , \tag{3.45}$$

where we have chosen $N_\psi = 1$.

In the following analysis we wish to consider our mechanism as the only source of both the dark matter and baryon asymmetry in the universe, and as such require that both Eqs. (3.1) and (3.3) are satisfied. This immediately leads to the following prediction for the dark matter mass in this scenario,

$$m_\psi \approx 116\epsilon \text{ GeV} . \tag{3.46}$$

Hence the dark matter candidate ψ must have a mass $m_\psi \approx 41$ GeV, or $m_\psi \approx 116$ GeV; for $T_{\text{rh}} > T_c$ and $T_{\text{rh}} < T_c$ respectively. For consistency with the initial assumption that $m_\psi \ll H_{\text{inf}}$, the inflationary Hubble rate must be greater than $\sim 10^3$ GeV. Interestingly, the fixing of the ratio ρ_D/ρ_B sets the mass of the associated dark fermion, and is independent of the $B - L$ charge of the dark matter candidate. If the mass of the dark matter candidate is lower than 116ϵ GeV, then it cannot be the only component of the dark matter energy density of the universe and an additional component to the dark matter must be introduced, but we do not consider that scenario here. It should be noted that this mass relation assumes the correct baryon asymmetry parameter is generated, and hence the other parameters of the model are constrained, which we consider now.

The required replication of the observed baryon asymmetry and dark to luminous mass density ratio results in the following condition on the model parameters, for $\mu = H_0$,

$$\eta_B \approx \epsilon |\mathcal{A}_1|^{5/2} \frac{101}{21 + |q_\psi|} \frac{1}{\pi^{13} g_*} \frac{m_X^5}{f_X^5} e^{-3N_e} \left(\frac{\kappa}{\mu T_{\text{rh}}^2} \right)^{\frac{3}{2}} \tag{3.47}$$

$$\approx 3.5 \times 10^{-18} \text{ GeV}^{-1/2} \, \epsilon \frac{|\mathcal{A}_1|^{5/2}}{21 + |q_\psi|} \frac{m_X^5}{f_X^5} \frac{H_{\text{inf}}}{T_{\text{rh}}^2} \left(\frac{\phi_0'}{f_X} \right)^{\frac{3}{2}} . \tag{3.48}$$

It is found that this can satisfy Eq. (3.1) for a wide range of parameter values. To see this we shall consider an example. Let us first assume ϕ_0' takes its maximal value, $H_{\text{inf}} M_p$, to ensure maximal $\mathcal{CP}$ violation. We will also identify the scale f_X with the inflationary Hubble rate, $\sim H_{\text{inf}}$, and the $B - L$ charge of the dark matter fermion to be $q_\psi = -1$. Under these assumptions we derive the following relation on the model parameters from Eqs. (3.1) and (3.48),

$$\eta_B \approx 7 \times 10^9 \text{ GeV } \epsilon g_X^5 \frac{H_{\text{inf}}}{T_{\text{rh}}^2} \quad \Rightarrow \quad \epsilon g_X^5 \frac{H_{\text{inf}}}{T_{\text{rh}}^2} \approx 10^{-20} \text{ GeV}^{-1} , \tag{3.49}$$

which reduces to:

$$g_X \sim 10^{-4} \text{ GeV}^{-1/5} \frac{T_{\text{rh}}^{2/5}}{H_{\text{inf}}^{1/5}} \quad \text{or} \quad m_X \sim 10^{-4} \text{ GeV}^{-1/5} T_{\text{rh}}^{2/5} H_{\text{inf}}^{4/5} . \tag{3.50}$$

In this example, the satisfaction of this relation guarantees the correct baryon asymmetry and hence relic dark matter density. This relation leads to interesting constraints on the allowed parameter space. We find that for instantaneous reheating, $H_{\text{inf}} \sim \frac{(T_{\text{rh}}^{\text{max}})^2}{M_p}$, we require $g_X \sim 0.4$. This is likely larger than that assumed in the model ($g_X \ll 1$), and means that this relation can be satisfied for almost all allowed reheating temperatures for a given inflationary Hubble rate. Hence, the reheating epoch must not be instantaneous and the reheating temperature cannot conflict with BBN constraints ($T_{\text{rh}} > \mathcal{O}(1)$ MeV). From the mass of the dark matter candidate and constraints on the inflationary potential we require that the inflationary Hubble rate is in the range 10^{14} GeV $> H_{\text{inf}} > 10^3$ GeV. Taking this into account, we see that the size of the coupling g_X that can satisfy this relation lies in the range $10^{-8} \lesssim g_X \lesssim 0.4$, with the lowest value corresponding to the input parameters $H_{\text{inf}} = 10^{14}$ GeV and $T_{\text{rh}} = \mathcal{O}(1)$ MeV. Interestingly, the mass of the gauge boson can be as low as $m_X \sim \mathcal{O}(1)$ MeV, when taking $H_{\text{inf}} = 10^3$ GeV and $T_{\text{rh}} = \mathcal{O}(1)$ MeV, leading to a coupling of order 10^{-6}.

It could be possible to probe experimentally the areas of parameter space in which the X boson mass is lowest, although this is made difficult by the associated tiny coupling constant. The generically small couplings would also make it difficult to detect the predicted dark matter candidate.

If we now consider the IR cut-off to be the Hubble rate at the beginning of BBN, $H_{\text{BBN}} \approx 10^{-25}$ GeV, the condition becomes,

$$\eta_B \approx \epsilon |A_1|^{5/2} \frac{101}{21 + |q_\psi|} \frac{1}{\pi^{13} g^*} \frac{m_X^5}{f_X^5} e^{-3N_e} \left(\frac{\kappa}{\mu T_{\text{rh}}^2} \right)^{\frac{3}{2}} \tag{3.51}$$

$$\approx 10^{-43} \text{ GeV}^{-1/2} \epsilon \frac{|A_1|^{5/2}}{21 + |q_\psi|} \frac{m_X^5}{f_X^5} \frac{H_{\text{inf}}}{T_{\text{rh}}^2} \left(\frac{\phi_0'}{f_X} \right)^{\frac{3}{2}} . \tag{3.52}$$

This choice of cut-off provides a significantly more constrained result, but can still be satisfied with an appropriate choice of parameters. Considering the same example model discussed above,

$$\eta_B \approx 2 \times 10^{-16} \text{ GeV } \epsilon g_X^5 \frac{H_{\text{inf}}}{T_{\text{rh}}^2} \quad \Rightarrow \quad \epsilon g_X^5 \frac{H_{\text{inf}}}{T_{\text{rh}}^2} \approx 4 \cdot 10^5 \text{ GeV}^{-1} , \tag{3.53}$$

which reduces to,

$$g_X \sim 10\,\text{GeV}^{-1/5}\frac{T_{\text{rh}}^{2/5}}{H_{\text{inf}}^{1/5}} \quad \text{or} \quad m_X \sim 10\,\text{GeV}^{-1/5}T_{\text{rh}}^{2/5}H_{\text{inf}}^{4/5} \,. \tag{3.54}$$

Considering instantaneous reheating requires a coupling of $g_X \sim 400$, this means that an extended reheating epoch is needed for consistency with our assumption that $g_X \ll 1$. The size of the couplings that can satisfy this relation lie in the range $10^{-3} \lesssim g_X \ll 1$, where the lower bound is for $H_{\text{inf}} = 10^{14}$ GeV and $T_{\text{rh}} = \mathcal{O}(1)$ MeV. This allowed window is much smaller than found in the previous case. The minimum mass of the boson is much higher with $m_X \sim 160$ GeV, when $H_{\text{inf}} = 10^3$ GeV and $T_{\text{rh}} = \mathcal{O}(1)$ MeV. Due to the higher couplings in this case, the model is likely much easier to constrain with experiments in the lower m_X regime, hence a very large H_{inf} and low T_{rh} would be required.

Case 2: $U(1)_B$ and a sterile baryon

If we now consider a gauged baryon number extension to the SM we must sum over all of the baryonic degrees of freedom, $\sum_i N_{SM}^i |q_{SM}^i| = 12$. In this scenario the contributions of the mixed anomalies $SU(2)^2 \times U(1)_B$ and $U(1)_Y^2 \times U(1)_B$ must be included, which generate a net charge only in the form of luminous matter. Hence, we find that the dark matter to luminous matter mass density ratio is given by,

$$\frac{\rho_D}{\rho_B} = \frac{1}{\epsilon}\frac{N_\psi}{12+N_\psi|q_\psi|}\frac{|\mathcal{A}_1|^{5/2}}{\frac{12|\mathcal{A}_1|^{5/2}}{N_\psi|q_\psi|+12}\frac{m_X^5}{f_X^5}+|\mathcal{A}_2|^{5/2}g_1^5+3|\mathcal{A}_3|^{5/2}g_2^5}\frac{m_X^5}{f_X^5}\left(\frac{m_\psi}{1\,\text{GeV}}\right) \tag{3.55}$$

$$\approx \frac{1}{\epsilon}\frac{|q_\psi|^{15/2}}{12|q_\psi|^{15/2}\frac{m_X^5}{f_X^5}+|q_\psi|+12}\frac{m_X^5}{f_X^5}\left(\frac{m_\psi}{1\,\text{GeV}}\right), \tag{3.56}$$

where $m_B = 1$ GeV and $q_B = 1$ have been set. In the second line we have taken $g_1^2 \simeq \frac{4\pi}{60}$ and $g_2^2 \simeq \frac{4\pi}{29}$, and used the anomaly values given above: $\mathcal{A}_2 = 3/2$ and $\mathcal{A}_3 = -3/2$. Upon rearranging, and requiring Eq. (3.3), we find the following expression for the mass of the dark matter candidate,

$$m_\psi \approx \epsilon\frac{f_X^5}{m_X^5}\frac{11(12|q_\psi|^{15/2}\frac{m_X^5}{f_X^5}+|q_\psi|+12)}{2|q_\psi|^{15/2}}\,\text{GeV}\,, \tag{3.57}$$

which simplifies to,

$$m_\psi \approx \frac{70}{\epsilon g_X^5}\,\text{GeV}\,, \tag{3.58}$$

when taking $|q_\psi| = 1$ and $g_X \ll 1$. Therefore, in this scenario we require that $H_{\text{inf}} \gtrsim 10^7$ GeV, which implies that the mass of the dark matter candidate lies in the range 10^6 GeV $\lesssim m_\psi \ll H_{\text{inf}}$. The required dark matter mass is greater than that found in the previous case due to the additional contributions to the luminous sector from the mixed anomalies, and the interplay of this with the mass density ratio.

The associated constraint imposed by the observed baryon asymmetry is given by,

$$\eta_B \approx \epsilon \frac{\mathcal{A}}{\pi^{13} g_*} e^{-3N_e} \left(\frac{\kappa}{\mu T_{rh}^2} \right)^{\frac{3}{2}} \tag{3.59}$$

$$\approx 3 \times 10^{-19} \text{ GeV}^{-1/2} \, \epsilon \mathcal{A} \frac{H_{inf}}{T_{rh}^2} \left(\frac{\phi_0'}{f_X} \right)^{\frac{3}{2}}, \tag{3.60}$$

where $\mathcal{A} = \left(\frac{12|q_\psi|^{15/2}}{|q_\psi|+12} \frac{m_X^5}{f_X^5} + |\mathcal{A}_2|^{5/2} g_1^5 + 3|\mathcal{A}_3|^{5/2} g_2^5 \right)$. As we assume $g_X^2 \ll 1$ this reduces to $\mathcal{A} \sim 1$, except when considering very large q_ψ, which we do not here. Similar to the $U(1)_{B-L}$ scenario, we assume ϕ_0' takes its maximal value, $H_{inf} M_p$, and identify the scale f_X with the inflationary Hubble rate, H_{inf}. Here the B charge of the dark matter fermion is chosen to be $q_\psi = 1$. Under these assumptions, we derive the following relation on the model parameters from Eqs. (3.1) and (3.48),

$$\eta_B \approx 4 \times 10^8 \text{ GeV } \epsilon \frac{H_{inf}}{T_{rh}^2} \rightarrow \epsilon \frac{H_{inf}}{T_{rh}^2} \approx 2 \cdot 10^{-19} \text{GeV}^{-1} \rightarrow H_{inf} \approx \frac{2 \cdot 10^{-19} \text{ GeV}^{-1}}{\epsilon} T_{rh}^2. \tag{3.61}$$

Interestingly, this relation is only dependent on the inflationary Hubble rate and the reheating temperature. Upon further inspection it appears to be almost consistent with instantaneous reheating, or a very efficient reheating process, hence in this scenario a close to maximal reheating temperature is required for any given inflationary Hubble rate.

From the analysis of the allowed dark matter mass range above, we find that the inflationary rate must be in the window 10^{14} GeV $\gtrsim H_{inf} \gtrsim 10^7$ GeV. The corresponding allowed range of reheating temperatures is then 10^{16} GeV $\gtrsim T_{rh} \gtrsim 7 \cdot 10^{12}$ GeV. The Eq. (3.58) and allowed mass range also give us a lower bound on the allowed coupling, which we find is $g_X^2 \sim 10^{-5}$, although this can be relaxed for larger dark matter charge values. These constraints also imply that the mass of the new X gauge boson must be at least greater than ~ 50 TeV, which suppresses the ability to detect the new boson and dark matter fermion. These values of the gauge coupling and boson mass range are well out of the range of any current collider experiments [81].

The constraint obtained when taking the IR cut-off to be $\mu = H_{BBN} \approx 10^{-25}$ GeV is,

$$\eta_B \approx 10^{-44} \text{ GeV}^{-1/2} \, \epsilon \mathcal{A} \frac{H_{inf}}{T_{rh}^2} \left(\frac{\phi_0'}{f_X} \right)^{\frac{3}{2}}. \tag{3.62}$$

Considering the same assumptions as for the previous cut-off we find,

$$\eta_B \approx 10^{-17} \text{ GeV } \epsilon \frac{H_{\text{inf}}}{T_{\text{rh}}^2} \rightarrow \epsilon \frac{H_{\text{inf}}}{T_{\text{rh}}^2} \approx 6 \cdot 10^6 \text{ GeV}^{-1} \rightarrow H_{\text{inf}} \approx \frac{6 \cdot 10^6 \text{ GeV}^{-1}}{\epsilon} T_{\text{rh}}^2 \,.$$

$$(3.63)$$

The constraints on the mass of the dark matter particle and coupling are the same for this case, but now the relation between the inflationary Hubble rate and reheating temperature has changed. Now we require an extended reheating epoch, with a maximum reheating temperature of $T_{\text{rh}} \sim 4$ TeV for $H_{\text{inf}} \sim 10^{14}$ GeV and minimum of $T_{\text{rh}} \sim 1$ GeV when $H_{\text{inf}} \sim 10^7$ GeV. This change in the allowed reheating temperatures is the only alteration caused by the use of the other cut-off.

In summary, the ability to detect the introduced gauge boson appears likely to be difficult due to the generic requirement of a high mass and small coupling, except in special cases. The dark matter candidate would be more easily detectable in the case of the gauged $B - L$ compared to the gauged B scenario due to it having a significantly smaller mass, which is of order $10 - 100$ GeV. Although as the dark matter candidate can only interact with the SM particles via the $U(1)_X$ gauge boson, the ability to detect it is highly dependent on the gauge boson mass and coupling, for which there is a wide range of possibilities. It may be possible to obtain constraints on the parameter space from Z' searches [68, 81–93].

3.7 Conclusions and Future Prospects

In this chapter, we have investigated a model for simultaneous generation of luminous and dark matter during the inflationary epoch, achieved through the introduction of an anomalous gauge interaction and sterile fermion to the SM. It has been found that this scenario for cogenesis can successfully reproduce observations for the two possible cases considered—a gauged B and a gauged $B - L$ charge.

In each scenario, we considered the parameter spaces that correctly predict both the dark matter to baryonic matter density ratio and baryon asymmetry. Interestingly, for the $U(1)_{B-L}$ extension we obtain a prediction for the mass of the dark matter candidate which is independent of the other choice of parameters, assuming a given relation between g_X, T_{rh} and H_{inf}. While in the $U(1)_B$ case, we find the model replicates the observed η_B as long as a relation between the reheating temperature and inflationary Hubble rate is adhered to, and the coupling satisfies $g_X^2 > 10^{-5}$.

The general mechanism for cogenesis developed here can be applied to more complex models involving other or extra anomalous gauge symmetries, as well as additional sterile or non-sterile fermionic states. It is possible that these additions could lead to a loosening of the constraints on the allowed parameters imposed by the observed matter-antimatter asymmetry, through extra contributions to the generation of luminous matter or dark matter. This would likely lead to altered requirements on the mass of the dark matter candidate.

Our general framework could also provide a mechanism for magnetogenesis, through the production of universal hypermagnetic fields via the $U_Y(1)$ CS term [94].

One could craft a model that not only generates this but also the particle asymmetries, providing an extra route for constraining the allowed parameter space.

Further to this, the study of the associated collider phenomenology of these models is of interest. Although, this is made difficult by the possibility of very small couplings and high masses. Despite this, there are certain areas of the parameter space that can already be constrained by results from collider experiments.

References

1. N.D. Barrie, A. Kobakhidze, Inflationary baryogenesis in a model with gauged baryon number. JHEP **09**, 163 (2014). https://doi.org/10.1007/JHEP09(2014)163
2. N.D. Barrie, A. Kobakhidze, Generating luminous and dark matter during inflation. Mod. Phys. Lett. A **32**(14), 1750087 (2017). https://doi.org/10.1142/S0217732317500870
3. A.G. Cohen, D.B. Kaplan, A.E. Nelson, Progress in electroweak baryogenesis. Ann. Rev. Nucl. Part. Sci. **43**, 27–70 (1993). https://doi.org/10.1146/annurev.ns.43.120193.000331
4. C.L. Bennett et al., Nine-year Wilkinson Microwave Anisotropy Probe (WMAP) observations: final maps and results. Astrophys. J. Suppl. **208**, 20 (2013). https://doi.org/10.1088/0067-0049/208/2/20
5. P.A.R. Ade et al., Planck 2013 results. I. Overview of products and scientific results. Astron. Astrophys. **571**, A1 (2014). https://doi.org/10.1051/0004-6361/201321529
6. P.A.R. Ade et al., Planck 2015 results XIII. Cosmological parameters. Astron. Astrophys. **594**, A13 (2016). https://doi.org/10.1051/0004-6361/201525830
7. J. Beringer et al., Review of Particle Physics (RPP). Phys. Rev. D **86**, 010001 (2012). https://doi.org/10.1103/PhysRevD.86.010001
8. G. Steigman, Primordial nucleosynthesis in the precision cosmology era. Ann. Rev. Nucl. Part. Sci. **57**, 463–491 (2007). https://doi.org/10.1146/annurev.nucl.56.080805.140437
9. V. Simha, G. Steigman, Constraining the early-universe baryon density and expansion rate. JCAP **0806**, 016 (2008). https://doi.org/10.1088/1475-7516/2008/06/016
10. G. Steigman, Primordial nucleosynthesis: the predicted and observed abundances and their consequences. PoS, NICXI:001 (2010)
11. B.D. Fields, P. Molaro, S. Sarkar, Big-Bang nucleosynthesis. Chin. Phys. C **38**, 339–344 (2014)
12. G. Krnjaic, Can the baryon asymmetry arise from initial conditions? Phys. Rev. D **96**(3), 035041 (2017). https://doi.org/10.1103/PhysRevD.96.035041
13. A.D. Dolgov, NonGUT baryogenesis. Phys. Rep. **222**, 309–386 (1992). https://doi.org/10.1016/0370-1573(92)90107-B
14. K. Funakubo, CP violation and baryogenesis at the electroweak phase transition. Prog. Theor. Phys. **96**, 475–520 (1996). https://doi.org/10.1143/PTP.96.475
15. M. Trodden, Electroweak baryogenesis. Rev. Mod. Phys. **71**, 1463–1500 (1999). https://doi.org/10.1103/RevModPhys.71.1463
16. A. Riotto, M. Trodden, Recent progress in baryogenesis. Ann. Rev. Nucl. Part. Sci. **49**, 35–75 (1999). https://doi.org/10.1146/annurev.nucl.49.1.35
17. M. Dine, A. Kusenko, The origin of the matter antimatter asymmetry. Rev. Mod. Phys. **76**, 1 (2003). https://doi.org/10.1103/RevModPhys.76.1
18. J.M. Cline, Baryogenesis, in *Les Houches Summer School—Session 86: Particle Physics and Cosmology: The Fabric of Spacetime Les Houches*, France, July 31–August 25, 2006
19. W. Buchmuller, Baryogenesis: 40 Years Later, in *Proceedings on 13th International Symposium on Particles, strings, and cosmology (PASCOS 2007)*, London, UK, 2–7 July 2007
20. S. Weinberg, *Cosmology* (Oxford University Press, Oxford, 2008), p 593. ISBN 9780198526827

21. M. Shaposhnikov, Baryogenesis. J. Phys. Conf. Ser. **171**, 012005 (2009). https://doi.org/10.1088/1742-6596/171/1/012005

22. P.F. Perez, New paradigm for baryon and lepton number violation. Phys. Rep. **597**, 1–30 (2015). https://doi.org/10.1016/j.physrep.2015.09.001

23. A.D. Dolgov, A.D. Linde, Baryon asymmetry in inflationary universe. Phys. Lett. B **116**, 329 (1982). https://doi.org/10.1016/0370-2693(82)90292-1

24. A.D. Sakharov, Violation of CP Invariance, C Asymmetry, and Baryon Asymmetry of the universe. Pisma Zh. Eksp. Teor. Fiz. **5**, 32–35 (1967). https://doi.org/10.1070/PU1991v034n05ABEH002497. [Usp. Fiz. Nauk 161, 61(1991)]

25. S. Weinberg, Baryon and lepton nonconserving processes. Phys. Rev. Lett. **43**, 1566–1570 (1979). https://doi.org/10.1103/PhysRevLett.43.1566

26. P. Nath, P.F. Perez, Proton stability in grand unified theories, in strings and in branes. Phys. Rep. **441**, 191–317 (2007). https://doi.org/10.1016/j.physrep.2007.02.010

27. I. Dorsner, P.F. Perez, How long could we live? Phys. Lett. B **625**, 88–95 (2005). https://doi.org/10.1016/j.physletb.2005.08.039

28. S. Dodelson, L.M. Widrow, Baryon symmetric baryogenesis. Phys. Rev. Lett. **64**, 340–343 (1990). https://doi.org/10.1103/PhysRevLett.64.340

29. S. Dodelson, L.M. Widrow, Baryogenesis in a baryon symmetric universe. Phys. Rev. D **42**, 326–342 (1990). https://doi.org/10.1103/PhysRevD.42.326

30. V.A. Kuzmin, A Simultaneous solution to baryogenesis and dark matter problems. Phys. Part. Nucl. **29**, 257–265 (1998). https://doi.org/10.1134/1.953070. [Phys. Atom. Nucl. 61,1107 (1998)]

31. K. Petraki, R.R. Volkas, Review of asymmetric dark matter. Int. J. Mod. Phys. A **28**, 1330028 (2013). https://doi.org/10.1142/S0217751X13300287

32. M.D. Schwartz, *Quantum Field Theory and the Standard Model* (Cambridge University Press, 2014). ISBN 1107034736, 9781107034730

33. L.H. Ryder, *Quantum Field Theory*. (Cambridge University Press, 1996). ISBN 9780521478144, 9781139632393, 9780521237642

34. T.P. Cheng, L.F. Li., *Gauge Theory of Elementary Particle Physics*. (Clarendon, Oxford Science Publications, Oxford, 1984), p. 536. ISBN 9780198519614

35. G.V. Dunne, Aspects of chern-simons theory, in *Topological Aspects of Low-dimensional Systems: Proceedings, Les Houches Summer School of Theoretical Physics, Session 69: Les Houches*, France, 7–31 July 1998

36. V.A. Rubakov, M.E. Shaposhnikov, Electroweak baryon number nonconservation in the early universe and in high-energy collisions. Usp. Fiz. Nauk **166**, 493–537 (1996). https://doi.org/10.1070/PU1996v039n05ABEH000145

37. E.J. Weinberg, Classical solutions in quantum field theories. Ann. Rev. Nucl. Part. Sci. **42**, 177–210 (1992)

38. P.B. Arnold, L.D. McLerran, Sphalerons, small fluctuations and baryon number violation in electroweak theory. Phys. Rev. D **36**, 581 (1987). https://doi.org/10.1103/PhysRevD.36.581

39. P.B. Arnold, An introduction to baryon violation in standard electroweak theory, in *Testing the Standard Model - TASI-90: Theoretical Advanced Study Inst. in Elementary Particle Physics Boulder*, Colorado, 3–29 June 1990, pp. 719–742

40. F.R. Klinkhamer, N.S. Manton, A saddle point solution in the weinberg-salam theory. Phys. Rev. D **30**, 2212 (1984). https://doi.org/10.1103/PhysRevD.30.2212

41. D. Diakonov, Instantons at work. Prog. Part. Nucl. Phys. **51**, 173–222 (2003). https://doi.org/10.1016/S0146-6410(03)90014-7

42. O. Espinosa, High-energy behavior of baryon and lepton number violating scattering amplitudes and breakdown of unitarity in the standard model. Nucl. Phys. B **343**, 310–340 (1990). https://doi.org/10.1016/0550-3213(90)90473-Q

43. M. Hellmund, J. Kripfganz, The decay of the sphalerons. Nucl. Phys. B **373**, 749–760 (1992). https://doi.org/10.1016/0550-3213(92)90274-F

44. M. Trodden, S.M. Carroll, TASI lectures: introduction to cosmology, in *Proceedings, Summer School, Progress in string theory (TASI 2003)*, Boulder, USA (2004) pp. 703–793, 2–27 June 2003

45. Y. Burnier, M. Laine, M. Shaposhnikov, Baryon and lepton number violation rates across the electroweak crossover. JCAP **0602**, 007 (2006). https://doi.org/10.1088/1475-7516/2006/02/007

46. L. Bento, Sphaleron relaxation temperatures. JCAP **0311**, 002 (2003). https://doi.org/10.1088/1475-7516/2003/11/002

47. S. Yu. Khlebnikov, M.E. Shaposhnikov, The statistical theory of anomalous fermion number nonconservation. Nucl. Phys. B **308**, 885–912 (1988). https://doi.org/10.1016/0550-3213(88)90133-2

48. R. Rangarajan, D.V. Nanopoulos, Inflationary baryogenesis. Phys. Rev. D **64**, 063511 (2001). https://doi.org/10.1103/PhysRevD.64.063511

49. S.H.-S. Alexander, M.E. Peskin, M.M. Sheikh-Jabbari, Leptogenesis from gravity waves in models of inflation. Phys. Rev. Lett. **96**, 081301 (2006). https://doi.org/10.1103/PhysRevLett.96.081301

50. S. Alexander, A. Marciano, D. Spergel, Chern-simons inflation and baryogenesis. JCAP **1304**, 046 (2013). https://doi.org/10.1088/1475-7516/2013/04/046

51. A. Maleknejad, M. Noorbala, M.M. Sheikh-Jabbari, Inflato-natural leptogenesis: leptogenesis in chromo-natural and gauge inflations (2012), arXiv:1208.2807

52. A. Maleknejad, Chiral gravity waves and leptogenesis in inflationary models with non-abelian gauge fields. Phys. Rev. D **90**(2), 023542 (2014). https://doi.org/10.1103/PhysRevD.90.023542

53. J. Preskill, Gauge anomalies in an effective field theory. Ann. Phys. **210**, 323–379 (1991). https://doi.org/10.1016/0003-4916(91)90046-B

54. M. Ibe, S. Matsumoto, T.T. Yanagida, The GeV-scale dark matter with B-L asymmetry. Phys. Lett. B **708**, 112–118 (2012). https://doi.org/10.1016/j.physletb.2012.01.032

55. S.M. Barr, The unification and cogeneration of dark matter and baryonic matter. Phys. Rev. D **85**, 013001 (2012). https://doi.org/10.1103/PhysRevD.85.013001

56. S. Kanemura, T. Matsui, H. Sugiyama, Neutrino mass and dark matter from gauged $U(1)_{B-L}$ breaking. Phys. Rev. D **90**, 013001 (2014). https://doi.org/10.1103/PhysRevD.90.013001

57. Z. Chacko, Y. Cui, S. Hong, T. Okui, Hidden dark matter sector, dark radiation, and the CMB. Phys. Rev. D **92**, 055033 (2015). https://doi.org/10.1103/PhysRevD.92.055033

58. A. Alves, A. Berlin, S. Profumo, F.S. Queiroz, Dirac-fermionic dark matter in U(1)$_X$ models. JHEP **10**, 076 (2015). https://doi.org/10.1007/JHEP10(2015)076

59. N. Okada, S. Okada, Z'-portal right-handed neutrino dark matter in the minimal U(1)$_X$ extended standard model. Phys. Rev. D **95**(3), 035025 (2017). https://doi.org/10.1103/PhysRevD.95.035025

60. W.-Z. Feng, P. Nath, Baryogenesis and dark matter in $U(1)$ extensions. Mod. Phys. Lett. A **32**, 1740005 (2017). https://doi.org/10.1142/S0217732317400053

61. B.A. Campbell, S. Davidson, J.R. Ellis, K.A. Olive, On the baryon, lepton flavor and right-handed electron asymmetries of the universe. Phys. Lett. B **297**, 118–124 (1992). https://doi.org/10.1016/0370-2693(92)91079-O

62. M. Giovannini, M.E. Shaposhnikov, Primordial hypermagnetic fields and triangle anomaly. Phys. Rev. D **57**, 2186–2206 (1998). https://doi.org/10.1103/PhysRevD.57.2186

63. M. Giovannini, Hypermagnetic knots, Chern-Simons waves and the baryon asymmetry. Phys. Rev. D **61**, 063502 (2000). https://doi.org/10.1103/PhysRevD.61.063502

64. R. Foot, G.C. Joshi, H. Lew, Gauged baryon and lepton numbers. Phys. Rev. D **40**, 2487–2489 (1989). https://doi.org/10.1103/PhysRevD.40.2487

65. C.D. Carone, H. Murayama, Possible light U(1) gauge boson coupled to baryon number. Phys. Rev. Lett. **74**, 3122–3125 (1995). https://doi.org/10.1103/PhysRevLett.74.3122

66. C.D. Carone, H. Murayama, Realistic models with a light U(1) gauge boson coupled to baryon number. Phys. Rev. D **52**, 484–493 (1995). https://doi.org/10.1103/PhysRevD.52.484

67. P.F. Perez, T. Han, T. Li, M.J. Ramsey-Musolf, Leptoquarks and neutrino masses at the LHC. Nucl. Phys. B **819**, 139–176 (2009). https://doi.org/10.1016/j.nuclphysb.2009.04.009

68. M. Duerr, P.F. Perez, M.B. Wise, Gauge theory for baryon and lepton numbers with leptoquarks. Phys. Rev. Lett. **110**, 231801 (2013). https://doi.org/10.1103/PhysRevLett.110.231801

69. T.R. Dulaney, P.F. Perez, M.B. Wise, Dark matter, baryon asymmetry, and spontaneous B and L breaking. Phys. Rev. D **83**, 023520 (2011). https://doi.org/10.1103/PhysRevD.83.023520

70. P.V. Dong, H.N. Long, A simple model of gauged lepton and baryon charges. Phys. Int. **6**(1), 23–32 (2010). https://doi.org/10.3844/pisp.2015.23.32

71. P.F. Perez, M.B. Wise, Baryon and lepton number as local gauge symmetries. Phys. Rev. D **82**, 011901 (2010). https://doi.org/10.1103/PhysRevD.82.079901, https://doi.org/10.1103/PhysRevD.82.011901

72. P.F. Perez, M.B. Wise, Low energy supersymmetry with baryon and lepton number gauged. Phys. Rev. D **84**, 055015 (2011). https://doi.org/10.1103/PhysRevD.84.055015

73. P.F. Perez, M.B. Wise, Breaking local baryon and lepton number at the TeV scale. *JHEP* **2011**(8), 68 (2011). https://doi.org/10.1007/JHEP08(2011)068

74. P. Schwaller, T.M.P. Tait, R. Vega-Morales, Dark matter and vectorlike leptons from gauged lepton number. Phys. Rev. D **88**(3), 035001 (2013). https://doi.org/10.1103/PhysRevD.88.035001

75. L. Parker, Particle creation in expanding universes. Phys. Rev. Lett. **21**, 562–564 (1968). https://doi.org/10.1103/PhysRevLett.21.562

76. L. Parker, Quantized fields and particle creation in expanding universes. I. Phys. Rev. **183**, 1057–1068 (1969). https://doi.org/10.1103/PhysRev.183.1057

77. O. Elgaroy, S. Hannestad, T. Haugboelle, Observational constraints on particle production during inflation. JCAP **0309**, 008 (2003). https://doi.org/10.1088/1475-7516/2003/09/008

78. N. Barnaby, Z. Huang, Particle production during inflation: observational constraints and signatures. Phys. Rev. D **80**, 126018 (2009). https://doi.org/10.1103/PhysRevD.80.126018

79. S.M. Carroll, *Spacetime and geometry: an introduction to general relativity* (Addison-Wesley, San Francisco, USA, 2004). ISBN 0805387323, 9780805387322

80. D.H. Lyth, The curvature perturbation in a box. JCAP **0712**, 016 (2007). https://doi.org/10.1088/1475-7516/2007/12/016

81. M. Carena, A. Daleo, B.A. Dobrescu, T.M.P. Tait, Z' gauge bosons at the Tevatron. Phys. Rev. D **70**, 093009 (2004). https://doi.org/10.1103/PhysRevD.70.093009

82. J.L. Rosner, Prominent decay modes of a leptophobic Z'. Phys. Lett. B **387**, 113–117 (1996). https://doi.org/10.1016/0370-2693(96)01022-2

83. H. Georgi, S.L. Glashow, Decays of a leptophobic gauge boson. Phys. Lett. B **387**, 341–345 (1996). https://doi.org/10.1016/0370-2693(96)00997-5

84. E. Accomando, A. Belyaev, L. Fedeli, S.F. King, C. Shepherd-Themistocleous, Z' physics with early LHC data. Phys. Rev. D **83**, 075012 (2011). https://doi.org/10.1103/PhysRevD.83.075012

85. P.J. Fox, J. Liu, D. Tucker-Smith, N. Weiner, An Effective Z'. Phys. Rev. D **84**, 115006 (2011). https://doi.org/10.1103/PhysRevD.84.115006

86. B.A. Dobrescu, F. Yu, Coupling-mass mapping of dijet peak searches. Phys.Rev. D. **88**(3), 035021 (2013). https://doi.org/10.1103/PhysRevD.88.035021, https://doi.org/10.1103/PhysRevD.90.079901

87. P.F. Perez, S. Ohmer, H.H. Patel, Minimal theory for lepto-baryons. Phys. Lett. B **735**, 283–287 (2014). https://doi.org/10.1016/j.physletb.2014.06.057

88. K.A. Olive et al., Review of particle physics. Chin. Phys. C **38**, 090001 (2014). https://doi.org/10.1088/1674-1137/38/9/090001

89. G. Aad et al., Search for new phenomena in the dijet mass distribution using $p-p$ collision data at $\sqrt{s} = 8$ TeV with the ATLAS detector. Phys. Rev. D **91**(5), 052007 (2015). https://doi.org/10.1103/PhysRevD.91.052007

90. G. Aad et al., Search for high-mass dilepton resonances in pp collisions at 8 TeV with the ATLAS detector. Phys. Rev. D **90**(5), 052005 (2014). https://doi.org/10.1103/PhysRevD.90.052005

91. J. Heeck, Unbroken B-L symmetry. Phys. Lett. B **739**, 256–262 (2014). https://doi.org/10.1016/j.physletb.2014.10.067

92. B.A. Dobrescu, Leptophobic boson signals with leptons, jets and missing energy (2015) arXiv:1506.04435

93. V. Khachatryan et al., Search for resonances and quantum black holes using dijet mass spectra in proton-proton collisions at $\sqrt{s} = 8$ TeV. Phys. Rev. D **91**(5), 052009 (2015). https://doi.org/10.1103/PhysRevD.91.052009
94. S.R. Zadeh, S.S. Gousheh, Effects of the $U_Y(1)$ Chern-Simons term and its baryonic contribution on matter asymmetries and hypermagnetic fields. Phys. Rev. D **95**(5), 056001 (2017). https://doi.org/10.1103/PhysRevD.95.056001

Chapter 4
Baryogenesis During Reheating via the Ratchet Mechanism

Another interesting and relatively unorthodox approach to generating the baryon asymmetry of the universe is via the inflaton during the reheating epoch. Unlike the dilution problems associated with inflationary cogenesis considered above, the difficulties with a reheating scenario are related to the high level of uncertainty and complexity of the dynamics associated with the reheating phase. The exact nature of the reheating epoch is mostly unknown, but it is a period dominated by the inflaton dynamics, and as such is strongly related to the properties of the inflaton and inflationary potential. To try to alleviate this issue and for simplicity, we will consider a Starobinsky inflationary scenario, which converges to the usual $\mu^2 \Phi^2$ potential during reheating.

We propose a new scenario for Baryogenesis during the reheating epoch that utilises the Ratchet mechanism, a model inspired by molecular motors in biological systems, and their ability to generate directed motion. This is achieved through the correlated behaviour between the inflaton and a complex scalar baryon. If the inflaton and the scalar baryon couple via a derivative coupling, the behaviour of the scalar baryon phase θ is found to be analogous to that of the forced pendulum, potentially producing a non-vanishing value of $\dot{\theta}$ which is necessary to generate a non-zero baryon number density [1].

4.1 The Reheating Epoch and Starobinsky Inflation

A consequence of the rapid expansion that occurs during inflation is that the energy densities of SM particles are diluted to negligible quantities, so in order to explain the observed matter energy densities today they must be produced dynamically some time after inflation ends. This phenomena is known as reheating, and involves the conversion of the energy density stored in the inflationary potential into matter, through the decay of the inflaton [2–12]. This process begins once the slow roll

© Springer International Publishing AG, part of Springer Nature 2018   89
N. D. Barrie, *Cosmological Implications of Quantum Anomalies*,
Springer Theses, https://doi.org/10.1007/978-3-319-94715-0_4

conditions are violated, that is, when the inflationary potential is no longer flat enough to support the inflationary scenario. In the canonical model, the almost homogeneous inflaton field oscillates coherently in its potential, being damped by Hubble expansion and a frictional term associated with the decay of the inflaton into matter particles and radiation. These processes are occurring in an out-of-equilibrium setting that continues until the decay rate becomes the dominant source of damping. For a $m^2\phi^2$ like inflaton potential during reheating, the coherent oscillations of the inflaton field induce an approximate matter dominated epoch [6]. Once the decay rate becomes the main source of damping, the universe will be dominated by relativistic particles produced from the distribution of the inflaton energy density into SM particles, and possibly other exotic species, and will begin to thermalise.

Once reheating is completed the universe thermalises and enters a radiation dominated epoch, with an initial temperature, also known as the reheating temperature, which is related to the initial energy in the inflationary potential and the decay rate of the inflaton. Faster inflaton decay leads to a more efficient transfer of energy from the inflationary sector to the radiation sector, and hence produces a higher reheating temperature. The exact details of the reheating epoch are complicated and mostly unknown, and as such it is an active area of scientific exploration, with possible probes of its properties considered [13–15]. We can constrain the maximal and minimal reheating temperatures using the maximum allowed inflationary scale and successful BBN, respectively, although this still permits a large range of reheating temperatures [16]. The uncertainty associated with this epoch lends itself to being an interesting source of answers to the mysteries surrounding cosmology and particle physics, particularly Baryogenesis [17], which we shall discuss in this chapter.

It is possible to determine a naive upper limit on the reheating temperature for a given inflationary mechanism by considering the Friedmann equations, and the approximate energy density contained within the inflaton potential,

$$\rho_{\text{inf}} \simeq U(\Phi_i) \simeq 3M_p^2 H_{\text{inf}}^2 \,, \tag{4.1}$$

where Φ_i is the initial value of the inflaton, $U(\Phi)$ is the inflaton potential, and $M_p = 1/\sqrt{8\pi G}$ is the reduced Planck mass. This energy density can then be equated to that associated with a radiation epoch with characteristic temperature T_{rh}, assuming efficient thermalisation [18], which is given by,

$$\rho_{\text{rh}} = \frac{\pi^2}{30} g_* T_{\text{rh}}^4 \,. \tag{4.2}$$

This gives a maximum reheating temperature of,

$$T_{\text{rh}} \simeq \left(\frac{90}{\pi^2 g_*}\right)^{1/4} \sqrt{M_p H_{\text{inf}}} \,. \tag{4.3}$$

For this temperature to be achieved, one must assume instantaneous and lossless decay of the inflaton into SM particles. The inflaton decay can be parametrised in the inflaton's equations of motion through the introduction of the decay width Γ, acting as an additional damping term,

$$\ddot{\Phi} + 3H\dot{\Phi} + \Gamma\dot{\Phi} + V'(\Phi) = 0 . \tag{4.4}$$

The value taken by Γ is dependent on the fields that the inflaton decays into and the strength of the associated interactions. The physical implications of the magnitude of Γ is that once $H \sim \Gamma$, the inflaton decays will dominate and reheating will end, leading to a radiation epoch with the characteristic temperature dictated by the properties of Γ. The length of the reheating epoch will be approximately Γ^{-1}. The associated reheating energy density can be found using the Born approximation,

$$\rho_{\text{rh}} \simeq 3\Gamma^2 M_p^2 , \tag{4.5}$$

which upon comparing to the energy density at the end of reheating, we find,

$$T_{\text{rh}} \simeq \left(\frac{90}{\pi^2 g_*}\right)^{1/4} \sqrt{\Gamma M_p} . \tag{4.6}$$

The number of e-folds of expansion that occur during reheating are,

$$N_{\text{rh}} = \frac{1}{3} \ln \left(\frac{\rho_{\text{inf}}}{\rho_{\text{rh}}}\right) \simeq \frac{2}{3} \ln \left(\frac{H_{\text{inf}}}{\Gamma}\right) . \tag{4.7}$$

It should be noted that the T_{rh} given in Eq. (4.6) is the maximum temperature of the radiation dominated universe, but it may not necessarily be the hottest temperature achieved after inflation has ended, rather, temperatures during reheating could be higher. This can have interesting implications for phase transitions and production of relics.

4.1.1 Starobinsky Inflation

Many models for the mechanism of inflation have been proposed, each of which must produce observational predictions consistent with the constraints depicted in Fig. 4.1. One model which is in good agreement with these constraints is the Starobinsky, or R^2, type models [19–21]. Given this agreement with the data, we shall consider this as the inflationary scenario in our investigation.

The Starobinsky action is the following,

$$S = \int dx^4 \sqrt{-g} \frac{M_p^2}{2} \left(R - \frac{R^2}{6\mu^2}\right) , \tag{4.8}$$

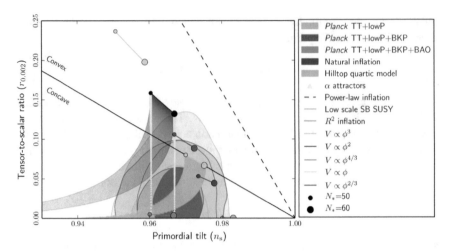

Fig. 4.1 Constraints on Inflationary observables from the PLANCK satellite, including predictions of various inflationary potentials [22]

where in the Einstein frame of the scalar parametrisation we have the following inflationary potential,

$$U(\Phi) = \frac{3\mu^2 M_p^2}{4}\left(1 - e^{-\sqrt{2/3}\Phi/M_p}\right)^2 ,\tag{4.9}$$

where $\mu = (1.3 \times 10^{-5})M_p$ is the inflaton mass [23]. From this potential the slow roll parameters and corresponding predictions can be found. Using the tools presented in Chap. 2, the Starobinsky model produces the following inflationary predictions,

$$n_s \simeq 1 - \frac{2}{N} \quad \text{and} \quad r \simeq \frac{12}{N^2} ,\tag{4.10}$$

where N is the number of inflationary e-folds.

The reheating period in the Starobinsky model approaches the usual $\frac{1}{2}\mu^2\Phi^2$, leading to this epoch being characterised as an approximate matter dominated epoch [6, 23, 24]. This can be seen in the expansion of the inflationary potential around Φ,

$$U(\Phi) = \frac{1}{2}\mu^2\Phi^2 - \frac{\mu^2}{\sqrt{6}M_p}\Phi^3 + \cdots\tag{4.11}$$

where the first term dominates for $\Phi < M_p$. From numerical calculations it is found that in the Starobinsky model the value the inflaton field takes at the beginning of the reheating epoch is [24],

$$\Phi_i = \Phi(t_i) = 0.62\, M_p ,\tag{4.12}$$

with a corresponding Hubble parameter of,

$$H_i = H(t_i) = 6.2 \times 10^{12} \text{ GeV} . \tag{4.13}$$

In the proceeding analysis, we will assume that the inflaton velocity $\dot{\Phi}(t_i)$ at this point in time is approximately zero for simplicity.

4.2 Baryogenesis During Reheating via the Ratchet Mechanism

Baryogenesis during reheating has been an area of interest in the literature for some time, given the complicated and uncertain nature of the epoch. In what follows we shall outline a new mechanism for Baryogenesis in which baryon number generation is driven by the oscillations of the inflaton field. As with any successful model of Baryogenesis that conserves CPT, the Sakharov conditions must be satisfied in order to reproduce the observed value of the baryon asymmetry parameter,

$$\eta_B =\simeq 8.5 \cdot 10^{-11} . \tag{4.14}$$

Of the many Baryogenesis models that have been proposed, there is a class that generates the baryon asymmetry via the coherent time-evolution of a complex scalar field that carries a baryon number charge [5, 17, 25–30]. In this work we consider a similar scenario in the setting of reheating, with a new scalar baryon coupled to the coherently oscillating inflaton field.

4.2.1 The Ratchet Mechanism

The new Baryogenesis mechanism we propose here acts during the reheating epoch, and is inspired by the ratchet models that describe molecular motors in biological systems [31, 32]. An example of this is the directed motion of myosin molecules along actin filaments which is achieved through cyclic chemical reactions that act as a driving force. The model we propose here utilises an analogous ratchet mechanism, in which the reflection symmetry of a scalar potential is broken by its interaction with the inflaton, with the inflaton's coherent oscillations providing a driving force. This driving force is supplied by the oscillation of the inflaton in its potential during reheating, while the position of the motor is embodied in the phase θ of the complex scalar baryon field. These two scalar fields interact via a derivative coupling which violates CP, breaking the reflection symmetry of the scalar baryon's potential. The form of the interaction between the two scalars is similar to that in the Baryogenesis mechanism considered in Ref. [33], although in their case CPT violation is required.

4.2.2 Description of the Model

We construct a model consisting of two scalar fields—a real scalar field Φ that we identify as the inflaton, and a complex scalar baryon ϕ. In the ensuing analysis we assume that the dynamics during reheating are dominated by these two scalars and only consider interactions of the inflaton with SM fields via an effective friction term. Our model is described by the following action,

$$
S = \int d^4x \sqrt{-g} \left[g_{\mu\nu} \partial^\mu \phi^* \partial^\nu \phi - V_0(\phi, \phi^*) \right.
$$
$$
\left. + \frac{1}{2} g_{\mu\nu} \partial^\mu \Phi \partial^\nu \Phi - U(\Phi) + \frac{i}{\Lambda} g_{\mu\nu} \left(\phi^* \overleftrightarrow{\partial^\mu} \phi \right) \partial^\nu \Phi \right],
$$
(4.15)

where $U(\Phi)$ is the inflationary potential, which we take to be the Starobinsky potential given in Eq. (4.9). The scalar baryon potential, $V_0(\phi, \phi^*)$, is defined as,

$$
V_0(\phi, \phi^*) = \lambda \phi^* \phi (\phi - \phi^*)(\phi^* - \phi) + \cdots ,
$$
(4.16)

where the ellipses denote terms that depend only on the product $\phi^* \phi$, which are not relevant to the dynamics in this mechanism.

The coupling between the inflaton and the scalar baryon is introduced as, $\mathcal{L}_{int} = -\frac{1}{\Lambda} j_B^\mu \partial_\mu \Phi = -\frac{\phi_r^2}{\Lambda} \partial^\mu \theta \partial_\mu \Phi$. This is a dimension five operator, and as such is suppressed by the mass scale Λ which is the energy cut-off at which the effective description of this scalar coupling breaks down. This interaction term also violates C and CP, which is a necessary ingredient for successful Baryogenesis.

Upon observation it can be seen that if $\lambda = 0$, the action will be invariant under the global $U(1)$ symmetry defined by the transformation $(\phi, \phi^*) \to (e^{i\alpha}\phi, e^{-i\alpha}\phi^*)$, where α is a constant. We identify the charge associated with this $U(1)$ symmetry as the baryon number B, where the complex scalar field ϕ is a baryon to which we assign unit baryonic charge. From the action we can calculate the corresponding baryon number current, or Noether current for the baryon symmetry, which is found to be,

$$
j_B^\mu = i(\phi \partial^\mu \phi^* - \phi^* \partial^\mu \phi) .
$$
(4.17)

We now consider the following polar coordinate parametrisation of the ϕ field,

$$
\phi = \frac{1}{\sqrt{2}} \phi_r e^{i\theta} .
$$
(4.18)

Under the global baryon number transformation, the phase θ transforms as $\theta \to \theta + \alpha$, while ϕ_r is invariant. In this parametrisation the baryon number density, which corresponds to the time component of Eq. (4.17), is given by,

$$n_B = j^0 = \phi_r^2 \dot{\theta} \, . \tag{4.19}$$

This implies that within the framework of our mechanism we must produce a non-zero $\dot{\theta}$, driven motion, to have a net baryon asymmetry generated. Rewriting the scalar baryon potential in terms of this reparametrisation, ϕ_r and θ, we obtain the following,

$$V(\phi_r, \theta) = V_0(\phi, \phi^*) = \lambda \phi_r^4 \sin^2 \theta + \cdots \tag{4.20}$$

where the ellipses now denote terms that depend only on ϕ_r. In the rest of our analysis, we assume that the terms that only depend on ϕ_r in $V(\phi_r, \theta)$ are such that they keep ϕ_r approximately fixed to a constant non-zero value, and that only the dynamics of the phase θ need be considered. The charge conjugation symmetry $\mathcal{C}$ is given by $\mathcal{C} : \phi \to \phi^*$, or $\theta \to -\theta$, so $\mathcal{C}$ is conserved in this potential. B invariance is related to the translational invariance of the phase θ, which is clearly violated by this potential, assuming $\lambda \neq 0$.

Therefore, in the new parametrisation of the scalar baryon the action takes the form,

$$S = \int d^4x \sqrt{-g} \left[\frac{\phi_r^2}{2} g_{\mu\nu} \partial^\mu \theta \partial^\nu \theta - \lambda \phi_r^4 \sin^2 \theta \right.$$
$$\left. + \frac{1}{2} g_{\mu\nu} \partial^\mu \Phi \partial^\nu \Phi - U(\Phi) - \frac{\phi_r^2}{\Lambda} g_{\mu\nu} \partial^\mu \theta \partial^\nu \Phi \right] . \tag{4.21}$$

Seeing as we wish to consider the cosmological setting of reheating we take the gravitational metric $g_{\mu\nu}$ to be the flat FRW metric,

$$ds^2 = g_{\mu\nu} dx^\mu dx^\nu = dt^2 - a^2(t) dx^2 \, . \tag{4.22}$$

Given this isotropic and homogeneous background, we extend this assumption to the properties of the scalar baryon and inflaton, for which spatial variation will be ignored in our analysis. In this background, the action of our model reads,

$$S = \int dt \, a(t)^3 \left[\frac{\phi_r^2}{2} \dot{\theta}^2 - \lambda \phi_r^4 \sin^2 \theta + \frac{1}{2} \dot{\Phi}^2 - U(\Phi) - \frac{\phi_r^2}{\Lambda} \dot{\theta} \dot{\Phi} \right] . \tag{4.23}$$

Therefore, in our model, the Sakharov conditions are satisfied in the following ways. Firstly, B violation is achieved by the scalar baryon potential $V(\phi_r, \theta)$. Secondly, the derivative coupling changes sign under both $\mathcal{C}$ and $\mathcal{CP}$. We have ignored the spatial dependence of the scalar fields, so the parity symmetry $\mathcal{P}$ is always conserved by this derivative coupling term. The required push out-of-equilibrium will be provided by the reheating epoch, induced by the coherent oscillation of the inflaton field. During reheating the derivative coupling between the scalar baryon and inflaton can lead to a kind of resonance effect when the respective potentials are of a similar order. As we shall see, when this is the case we can observe directed motion in the scalar baryon phase, and hence a non-zero baryon number density.

Now, we can carry out the calculations required to determine the baryon number density generated in this framework. In our analysis we will take the initial phase of the scalar baryon to be zero, placing the scalar baryon initially in the minimum of its potential. To ensure there is no initial bias between matter and antimatter we assume that the initial phase velocity $\dot{\theta}$ is zero.

4.3 Analytical Evaluation

We shall now find an analytical solution for the scalar baryon phase equation of motion so that we can determine the region of parameter space where we obtain driven motion, and hence have a non-zero baryon number density. Firstly, we find the equations of motion for θ and Φ, using the action presented in Eq. (4.23),

$$(\ddot{\Phi} + 3H\dot{\Phi}) + \left(\Gamma\dot{\Phi} + \frac{dU(\Phi)}{d\Phi}\right) - \frac{\phi_r^2}{\Lambda}(\ddot{\theta} + 3H\dot{\theta}) = 0 \,, \qquad (4.24)$$

$$(\ddot{\theta} + 3H\dot{\theta}) - \frac{1}{\Lambda}(\ddot{\Phi} + 3H\dot{\Phi}) = 0 \,, \qquad (4.25)$$

where $\Gamma\dot{\Phi}$ is the inflaton friction term, added in by hand, which encapsulates the decay of the inflaton into SM or mediator particles during reheating, and H is the Hubble parameter. After some rearrangement, the above equations read,

$$\left(1 - \frac{\phi_r^2}{\Lambda^2}\right)(\ddot{\Phi} + 3H\dot{\Phi}) + \left(\Gamma\dot{\Phi} + \frac{dU(\Phi)}{d\Phi}\right) + \frac{\lambda\phi_r^4}{\Lambda}\sin(2\theta) = 0 \,, \quad (4.26)$$

$$\left(1 - \frac{\phi_r^2}{\Lambda^2}\right)(\ddot{\theta} + 3H\dot{\theta}) + \frac{1}{\Lambda}\left(\Gamma\dot{\Phi} + \frac{dU(\Phi)}{d\Phi}\right) + \lambda\phi_r^2\sin(2\theta) = 0 \,. \quad (4.27)$$

The $\left(1 - \frac{\phi_r^2}{\Lambda^2}\right)$ factor can be absorbed through rescaling of the parameters Γ, μ^2 in $U(\Phi)$, and λ, as follows,

$$\tilde{\Gamma} \equiv \frac{\Gamma}{1 - \phi_r^2/\Lambda^2}, \qquad \tilde{\mu}^2 \equiv \frac{\mu^2}{1 - \phi_r^2/\Lambda^2}, \qquad \tilde{\lambda} \equiv \frac{\lambda}{1 - \phi_r^2/\Lambda^2} \,, \qquad (4.28)$$

and hence the equations simplify to,

$$(\ddot{\Phi} + 3H\dot{\Phi}) + \left(\tilde{\Gamma}\dot{\Phi} + \frac{d\tilde{U}(\Phi)}{d\Phi}\right) + \frac{\tilde{\lambda}\phi_r^4}{\Lambda}\sin(2\theta) = 0 \,, \tag{4.29}$$

$$(\ddot{\theta} + 3H\dot{\theta}) + \frac{1}{\Lambda}\left(\tilde{\Gamma}\dot{\Phi} + \frac{d\tilde{U}(\Phi)}{d\Phi}\right) + \tilde{\lambda}\phi_r^2\sin(2\theta) = 0 \,. \tag{4.30}$$

During reheating in the Starobinsky inflationary model, the oscillation of the massive inflaton gives rise to an approximate matter dominated epoch, during which the Hubble parameter H is approximately given by,

$$H = \frac{\dot{a}}{a} \approx \frac{2}{3t} \,. \tag{4.31}$$

We shall also assume that during reheating the Starobinsky potential can be expressed approximately as,

$$\tilde{U}(\Phi) \approx \frac{1}{2}\tilde{\mu}^2\Phi^2 \quad \rightarrow \quad \frac{d\tilde{U}(\Phi)}{d\Phi} \approx \tilde{\mu}^2\Phi \,, \tag{4.32}$$

as found in the expansion given in Eq. (4.11). Note that early on in the reheating epoch the full potential should be used. This approximation is valid towards the end of the epoch, when the inflaton is oscillating near the potential minimum. The equations of motion are now,

$$\left(\ddot{\Phi} + \frac{2}{t}\dot{\Phi}\right) + \left(\tilde{\Gamma}\dot{\Phi} + \tilde{\mu}^2\Phi\right) + \frac{\tilde{\lambda}\phi_r^4}{\Lambda}\sin(2\theta) = 0 \,, \tag{4.33}$$

$$\left(\ddot{\theta} + \frac{2}{t}\dot{\theta}\right) + \frac{1}{\Lambda}\left(\tilde{\Gamma}\dot{\Phi} + \tilde{\mu}^2\Phi\right) + \tilde{\lambda}\phi_r^2\sin(2\theta) = 0 \,. \tag{4.34}$$

In this form we can now begin solving these equations.

Behaviour of the Inflaton

We wish for the inflaton's motion to be unaffected by the dynamics of θ. This ensures that the properties of the reheating epoch are retained and the approximation given in Eq. (4.31) remains valid. To do so, we assume that the $\sin(2\theta)$ term in Eq. (4.33) can be neglected—the condition for this assumption shall be provided below. The equation of motion of the inflaton Φ then becomes,

$$\ddot{\Phi} + \left(\frac{2}{t} + \tilde{\Gamma}\right)\dot{\Phi} + \tilde{\mu}^2\Phi = 0 \,. \tag{4.35}$$

This equation can be easily solved in the case when $\tilde{\Gamma} \ll \tilde{\mu}$, which is a valid assumption in our scenario. The approximate solution to this equation is,

$$\Phi(t) = \Phi_i \left(\frac{t_i}{t} \right) e^{-\tilde{\Gamma}(t-t_i)/2} \cos\left[\tilde{\mu}(t - t_i) \right], \tag{4.36}$$

where t_i is the time at which the reheating epoch begins, and $\Phi_i = \Phi(t_i)$. This solution indicates that the motion of $\Phi(t)$ is oscillatory, with an angular frequency $\tilde{\mu}$, and is damped by $e^{-\tilde{\Gamma}t/2}$, coming from the $\tilde{\Gamma}$ term in the equation of motion, while the amplitude is also attenuated as a function of $1/t$ from the Hubble damping term.

Now that we have this solution it is possible to find a simple relation describing the assumption that the $\sin(2\theta)$ term can be neglected in the equation of motion of Φ. This requires the following relation be satisfied,

$$\frac{\tilde{\lambda}\phi_r^4}{\Lambda} \ll \text{amplitude of } \tilde{\mu}^2 \Phi(t). \tag{4.37}$$

This should be true throughout the reheating epoch, but is sufficient to be true at the end of reheating due to the damping of the inflaton's amplitude, predominantly by Hubble damping,

$$\frac{\tilde{\lambda}\phi_r^4}{\Lambda} \ll \tilde{\mu}^2 \Phi_i \left(\frac{H_f}{H_i} \right), \tag{4.38}$$

where t_f is the time at the end of the reheating epoch, and H_i and H_f are respectively the values of the Hubble parameter at t_i and t_f.

Behaviour of the Scalar Baryon

Now that we have determined the dynamics of the inflaton during reheating, we can utilise the solution to find an analytical solution for the phase of the scalar baryon. As found above, the equation of motion for θ is,

$$\left(\ddot{\theta} + \frac{2}{t}\dot{\theta} \right) + \tilde{\lambda}\phi_r^2 \sin(2\theta) + \frac{1}{\Lambda} \left[\tilde{\Gamma}\dot{\Phi}(t) + \tilde{\mu}^2\Phi(t) \right] = 0. \tag{4.39}$$

Since the amplitude of $\tilde{\Gamma}\dot{\Phi}(t)$ is suppressed compared to the amplitude of $\tilde{\mu}^2\Phi(t)$ by a factor of approximately $\tilde{\Gamma}/\tilde{\mu}$, we can drop the $\tilde{\Gamma}\dot{\Phi}(t)$ term. Doing this, and defining some new parameters the equation of motion for θ becomes,

$$\left(\ddot{\theta} + \frac{2}{t}\dot{\theta} \right) + p \sin(2\theta) + q(t) \cos\left[\tilde{\mu}(t - t_i) \right] = 0, \tag{4.40}$$

where we have defined,

$$p = \tilde{\lambda}\phi_r^2, \quad q(t) = \frac{\tilde{\mu}^2\Phi_i}{\tilde{\Lambda}} \left(\frac{t_i}{t} \right) e^{-\tilde{\Gamma}(t-t_i)/2}. \tag{4.41}$$

This equation is not simple to solve, so we shall first consider a few possible scenarios to determine some of its properties. First, consider the case when $p \ll q(t)$, we now have,

$$\left(\ddot{\theta} + \frac{2}{t}\dot{\theta}\right) = \frac{1}{t^2}\frac{d}{dt}\left(t^2\dot{\theta}\right) = -q(t)\cos\left[\tilde{\mu}(t - t_i)\right],\tag{4.42}$$

which can be integrated to yield,

$$\dot{\theta}(t) = \left(\frac{\tilde{\mu}^2\Phi_i}{\tilde{\Lambda}}\right)\frac{t_i}{t^2}\left[1 - e^{-\tilde{\Gamma}(t-t_i)/2}\left\{\cos\left[\tilde{\mu}(t - t_i)\right] + \tilde{\mu}t\sin\left[\tilde{\mu}(t - t_i)\right]\right\}\right] + \cdots\tag{4.43}$$

where the ellipses includes terms subleading in $\tilde{\Gamma}/\tilde{\mu}$, which we have dropped for consistency with our approximations above. Upon observation we find that there is no dependence on θ, specifically the $\sin(2\theta)$ term is absent. Thus, we can see that, in this limit the motion of $\dot{\theta}$ is driven solely by the oscillation of the inflaton and simply oscillates around zero, not maintaining any finite value. This is to be expected since this limit is equivalent to removing the B violating term, associated with the scalar baryon potential, from the equation.

Now consider the limit $p \gg q(t)$, for which the equation of motion becomes,

$$\ddot{\theta} + \frac{2}{t}\dot{\theta} + p\sin(2\theta) = 0 \,.\tag{4.44}$$

In this case, if we start from a state with finite energy, the friction term will damp the motion of the phase θ until it settles into one of its potential minima, and again there will be no non-zero $\dot{\theta}$ which persists and can lead to a non-zero baryon number density. Of course, this is to be expected since in this limit the C and CP breaking term has been removed.

Therefore, we can conclude that for successful Baryogenesis, we require $p \simeq q(t)$ so that both the B breaking and the C and CP terms can contribute to the time evolution of θ. In particular, we need the condition $p \simeq q(t)$ to be satisfied towards the end of the reheating epoch, since if it is satisfied too early, $q(t)$ will evolve to be smaller than p by the end of the epoch and any directed motion of θ will come to a stop. Thus, we require $p \simeq q(t_f)$, which we shall name the Sweet Spot Condition (SSC),

$$\tilde{\lambda}\phi_r^2 \simeq \frac{\tilde{\mu}^2\Phi_i}{\Lambda}\left(\frac{H_f}{H_i}\right) \,.\tag{4.45}$$

The SSC must be compatible with the prior assumptions we have made, such as that derived in Eq. (4.38). Upon combining these two conditions we find the following requirement, $\phi_r^2/\Lambda^2 \ll 1$, This can easily be satisfied, and has the flow on effect that we can drop the tilde notation from this point on, that is, $\tilde{\lambda} \approx \lambda$, $\tilde{\mu} \approx \mu$, and $\tilde{\Gamma} \approx \Gamma$.

4.3.1 Approximate Analytical Solution

Considering the SSC found in the previous section, the equation of motion is still difficult to solve, so we will first attempt to find an approximate analytical solution. We shall then compare this with numerical calculations to test for consistency.

The requirement for non-zero driven motion near the end of reheating is defined by the SSC, which can be written as follows,

$$\lambda \phi_r^2 \simeq \frac{H_f}{H_i} \frac{\mu^2 \Phi_i}{\Lambda} \approx 7 \times 10^{14} \text{ GeV} \frac{T_{\text{rh}}^2}{\Lambda} , \tag{4.46}$$

where we have used Eqs. (4.3), (4.12), (4.13) and (4.45). If we substitute the SSC into the equation of motion for θ we get,

$$\frac{9}{4} H^4 \theta'' - \frac{3}{2\Lambda} H^2 \Gamma \Phi' + \frac{\mu^2}{\Lambda} \frac{H}{H_i} \Phi_0 \cos \left[\frac{2\mu}{3H} \right] - \lambda \phi_r^2 \sin(2\theta) \approx 0 , \tag{4.47}$$

where the primes denote derivatives with respect to the Hubble rate. We obtain the following simplified equation,

$$\frac{9}{4} H^4 \theta'' \approx p \left(\sin(2\theta) - \cos \left[\frac{2\mu}{3H} \right] \right) , \tag{4.48}$$

where we have fixed $p = \lambda \phi_r^2 \simeq \frac{H_f}{H_i} \frac{\Phi_0 \mu^2}{\Lambda}$ near the end of reheating, the damping term has been assumed to be suppressed, and the reheating epoch has been taken to be sufficiently long such that $H_f \ll H_i$. In order to aid finding an approximate analytical solution we have neglected the Hubble damping term. In doing this we find the following simplified expression for the dynamics of θ near the end of reheating,

$$\ddot{\theta} \approx p \left(\sin(2\theta) - \cos(\mu t) \right) . \tag{4.49}$$

If we now assume the two oscillatory parts have different frequencies, with the inflaton oscillation rate μ being slower, we find that the average value of $\dot{\theta}$ is of order,

$$|\dot{\theta}| \approx \frac{p}{\mu} = \frac{\lambda \phi_r^2}{\mu} . \tag{4.50}$$

Therefore, the baryon number density generated at the end of reheating is given by,

$$n_{\text{B}} \approx \frac{\lambda \phi_r^4}{\mu} . \tag{4.51}$$

Combining the entropy density at the end of reheating and the approximate analytical expression for n_B derived in Eq. (4.51), we obtain the following expression for the asymmetry parameter,

$$\eta_B^{reh} \approx 0.02 \frac{\lambda \phi_r^4}{\mu T_{\rm rh}^3} \approx 0.1 \frac{\phi_r^2}{\Lambda T_{\rm rh}} \,, \tag{4.52}$$

where we have used the SSC in Eq. (4.46). This result assumes certain constraints on the parameters, namely the SSC, $\frac{\phi_r^2}{\Lambda^2} \ll 1$, and that the energy density of the scalar baryon must be less than that of the radiation epoch at the end of reheating, which corresponds to,

$$\lambda \phi_r^4 < \frac{\pi^2}{30} g_* T_{\rm rh}^4 \,. \tag{4.53}$$

It is found that a wide range of parameter choices can be used that satisfy the observed baryon asymmetry. We also find that when undertaking numerical calculations of Eq. (4.39), they are in reasonable agreement with this solution, with results being consistent to within an order of magnitude for most parameter choices where driven motion is observed. It is generically found that high reheating temperatures are required, that is $T_{rh} > 10^{13}$ GeV.

4.3.2 Phase Locked States

A more rigorous solution for θ can be found by drawing an analogy between our mechanism and a forced pendulum. The equation of motion for θ can be parametrised as follows,

$$\ddot{\theta} + f(t)\dot{\theta} + p \sin(2\theta) = -q(t) \cos[\mu(t - t_i)] \,, \tag{4.54}$$

where

$$p = \lambda \phi_r^2, \quad f(t) = \frac{2}{t}, \quad q(t) = \frac{\mu^2 \Phi_i}{\Lambda} \left(\frac{t_i}{t}\right) e^{-\Gamma(t-t_i)/2} \,. \tag{4.55}$$

Now we can observe that the motion of θ is identical to that of a forced pendulum. The term proportional to $\sin(2\theta)$ can be viewed as the gravitational force on the pendulum, when it is at an angle 2θ from the vertical, q the external pushing force, and f the friction term. There is an added complexity in our case, in that the strength of the external force $q(t)$ and the friction $f(t)$ on the pendulum depend on t.

As discussed above, Baryogenesis is realised in this scenario when the solution of the equations of motion is found to give $\dot{\theta}(t_f) \neq 0$, at the end of reheating t_f. This means we must adjust the timing and intensity of the external pushing to match the motion of the pendulum, which is the idea embodied by the SSC, corresponding to $p \approx q(t_f)$. If this is satisfied, the rotational motion of the pendulum around the fixed point arises with an almost constant angular velocity $\dot{\theta}$. This is the solution we wish to determine.

The time evolution of $q(t)$ toward the end of the reheating epoch can be expected to be slow, so to analyse the dynamics of θ within that time frame, it is sufficient to replace it with the constant $q(t_f)$. The same can be said of the coefficient of the $\dot{\theta}$ term, which we replace with $(2/t_f)$. Thus, we obtain the following simplified equation,

$$\ddot{\theta} + \frac{2}{t_f}\dot{\theta} + p\sin(2\theta) = -q(t_f)\cos[\mu(t - t_i)], \qquad (4.56)$$

where the coefficients of $\dot{\theta}$ and $\cos[\mu(t - t_i)]$ are now constants, $p = \lambda\phi_r^2$ and $q(t_f) = e^{-\frac{1}{3}\frac{H_{end}}{H_{ini}}}\frac{\Phi_0\mu^2}{\Lambda}$, the damping term has been taken to be suppressed, and $H_{end} \ll H_i$ has been assumed.

The relevant solutions to the above equation in our Baryogenesis scenario are those that increase or decrease monotonously in time with only small amplitude modulations. Such solutions exist and are known as phase-locked states, which are found in the study of the chaotic behaviour of the forced pendulum. The conditions for phase-locked states to exist were considered in the study of chaotic behaviour of electric current passing through a Josephson junction [34]. Since it is convenient to follow the notation adopted in these studies [35], we change the variables as follows,

$$\Theta \equiv 2\theta, \quad \tau \equiv \sqrt{2p}\left[(t - t_i) - \frac{\pi}{\mu}\right], \quad \omega \equiv \frac{\mu}{\sqrt{2p}}, \quad Q \equiv \sqrt{\frac{p}{2}}t_f, \quad \gamma \equiv \frac{q(t_f)}{p}. \qquad (4.57)$$

Thus, the equation of motion becomes,

$$\ddot{\Theta} + \frac{1}{Q}\dot{\Theta} + \sin\Theta = \gamma\cos(\omega\tau). \qquad (4.58)$$

Our equation coincides exactly with that of the forced pendulum or Josephson junctions. The generic phase-locked state solution to the above equation has the following form, when $\gamma \approx 1$,

$$\Theta(\tau) = \Theta_0 + n\omega\tau - \sum_{m=1}^{\infty}\alpha_m\sin(m\omega\tau - \phi_m), \qquad (4.59)$$

where n and m are integers. In the numerical calculations we performed only the phase-locked states with $m = 1$ appear. In such solutions the period of the amplitude modulation is equal to that of inflaton's oscillation. Hence the solution to our equation of motion is of the form,

$$\Theta = \Theta_n + n(\omega\tau - \phi) - \alpha\sin(\omega\tau - \phi). \qquad (4.60)$$

For these solutions, we can calculate the baryon number density n_B as the time average of $\dot{\Theta}$. From this we arrive at the following,

$$n_B = \phi_r^2 \langle \dot{\theta} \rangle = \sqrt{\frac{p}{2}} \phi_r^2 \langle \dot{\Theta} \rangle = \sqrt{\frac{p}{2}} \phi_r^2 n \omega = \left(\mu \phi_r^2 \right) \frac{n}{2} . \qquad (4.61)$$

Interestingly, this result depends on the integer n, where $n/2$ is the number of rotations of the phase θ per oscillation of the inflaton. The value of n depends on the validity of the phase-locked state and its stability. This is not a number which is predicted by the theory and hence we must determine it using numerical simulations. Although, the approximate solution derived in the previous section can provide an approximate value for n, which is found to be surprisingly consistent with numerical calculations.

4.4 The Generated Asymmetry Parameter

Combining the entropy density at the end of reheating and the analytical expression for n_B derived in Eq. (4.61), we obtain the following relation for the asymmetry parameter,

$$\eta_B^{\text{reh}} = \frac{n_B}{s} \approx 0.01 n \times \left(\frac{\mu \phi_r^2}{T_{\text{rh}}^3} \right) . \qquad (4.62)$$

This relation needs to be consistent with the observational constraint presented in Eq. (4.14). At the end of reheating, there is no further generation of baryon number, but for reheating temperatures greater than ~ 100 GeV, sphaleron redistribution must be considered [36–38]. This redistribution leads to a dilution factor of $\frac{28}{79}$, and hence the required asymmetry parameter at the end of reheating is,

$$\eta_B^{\text{reh}} = \frac{79}{28} \eta_B \simeq 2.4 \times 10^{-10} . \qquad (4.63)$$

The rough analytical expression that we derived, is found to be a good approximation for large values of n. By comparing these two solutions we can get an approximation of the parameter n. Combining Eqs. (4.52) and (4.62) we obtain,

$$n \approx 2 \frac{\lambda \phi_r^2}{\mu^2} , \qquad (4.64)$$

which must be greater than one if driven motion is to be observed. For values of n of $\mathcal{O}(10)$ and greater, the approximate solution is in good agreement with numerical simulations, which is likely due to the consistency of these n values with the assumptions used. In this mechanism, large values of λ are naturally required to produce driven motion, which we shall address in future work. We also require high reheating temperatures for the numerical simulations to be valid, although this can also be seen when substituting the SSC into Eq. (4.64),

$$n \approx 2 \times 10^{-12} \text{ GeV}^{-1} \frac{T_{\text{rh}}^2}{\Lambda} \,, \tag{4.65}$$

where a large reheating temperature is required for driven motion. It would generally be expected that the cutoff scale Λ would be of order of the Planck scale, such that the derivative coupling interaction term is valid near the beginning of reheating.

4.4.1 Conversion to Standard Model Particles

So far we have identified the baryon number density n_B with that of the scalar baryon, implying $n_B(\text{scalar}) = \phi_r^2 \dot{\theta}$. However, the actual baryon number density in our universe is made up of fermionic matter. Therefore, the baryon number density of the scalar baryon must be converted to that of fermionic baryons. There are multiple ways to accomplish this conversion.

If we insist on using SM fields directly for the scalar-fermion conversion, we require $SU(3)_C \times SU(2)_L \times U(1)_Y \times B$ invariant interactions.

The simplest and most realistic model may be to reinterpret the baryon number associated with the scalar as lepton number, and introduce the lepton-number preserving dimension four interaction,

$$\Delta \mathcal{L}_{\text{int}} = y_L \phi^* \bar{\nu}_R^c \nu_R + \text{h.c.} \tag{4.66}$$

which describes the decay of the complex scalar lepton into a $\nu_R \nu_R$ pair. This same interaction can be used to generate a large Majorana mass for ν_R, when ϕ obtains the vacuum expectation value, $\langle \phi \rangle = \phi_r/\sqrt{2}$, with this leptonic scalar being a component of the neutrino mass generating model known as the seesaw mechanism [39–41]. The lepton number generated via the decay of the complex scalar can then be converted to baryon number through redistribution by $B - L$ conserving sphaleron processes [42, 43], as in the usual Leptogenesis scenarios [44].

Alternatively, one could introduce the following dimension six interaction between the scalar and fermionic baryon number currents,

$$\Delta \mathcal{L}_{\text{int}} = \frac{i}{\Lambda^2} \left(\phi^* \overleftrightarrow{\partial}_\mu \phi \right) \frac{1}{3} \left(\bar{u} \gamma^\mu u + \bar{d} \gamma^\mu d \right) , \tag{4.67}$$

where u and d are four component Dirac fields. Identifying the charges carried by the scalar and fermionic currents requires the existence of a term in the interaction Lagrangian which would lead to both the ϕ and the quark fields transforming at the same time, which we will not show explicitly here. Rewriting the scalar baryon in the polar coordinate parametrisation, $\phi = \frac{1}{\sqrt{2}} \phi_r e^{i\theta}$, and ignoring the spatial dependence of θ we obtain,

$$\Delta \mathcal{L}_{\text{int}} = -\frac{\phi_r^2}{\Lambda^2} \dot{\theta} \frac{1}{3} \left(u^\dagger u + d^\dagger d \right) , \tag{4.68}$$

such an interaction term has been utilised in previously considered Baryogenesis scenarios [33]. This term is analogous to a chemical potential coupling to the baryonic current, shifting in favour of matter or antimatter, where $\mu_B = -\frac{\phi_r^2}{\Lambda^2}$.

4.5 Conclusions and Future Prospects

In this chapter, we have considered a new Baryogenesis mechanism that acts during reheating, which takes its inspiration from the ratchet models of molecular motors in biological systems. The mechanism we propose here is found to produce driven motion in an analogous framework to that found for a forced pendulum, in which the driving force is supplied by the oscillation of the inflaton, the position of the motor is embodied in the phase θ of a complex scalar baryon field, and the required breaking of the reflection symmetry is realised via the coupling of the inflaton to the scalar baryon. The push out-of-equilibrium comes from the reheating epoch itself. In our analysis, we find a rigorous solution which is dependent on an indeterminable parameter n,

$$\eta_B^{reh} \approx 0.01n \times \left(\frac{\mu\phi_r^2}{T_{rh}^3} \right) , \qquad (4.69)$$

The issue with this solution is that the parameter n must be determined using numerical calculations. This can be bypassed by considering the approximate solution we derived,

$$\eta_B^{reh} \approx 0.02 \frac{\lambda\phi_r^4}{\mu T_{rh}^3} \approx 0.1 \frac{\phi_r^2}{\Lambda T_{rh}} , \qquad (4.70)$$

which is in good agreement with numerical simulations for $n > 10$. Using these relations we found that it is possible to replicate the observed baryon asymmetry. We were also able to find an estimate of the parameter n through comparing the approximate and rigorous solutions, which allows a more targeted approach at testing the allowed parameter space with the numerical methods.

Although, in order for this to be achieved a very unnatural choice of the coupling λ must be chosen. This issue can be alleviated by a change in the scalar baryon potential we have been exploring, which shall be discussed in future work.

High reheating temperatures are a generic requirement of our model, which can be seen in the SSC, but is also a result of difficulties with the numerical calculations when considering reheating temperatures less than approximately 5×10^{13} GeV. Reheating temperatures greater than 10^{14} GeV could be a possible issue due to the over production of gravitinos, if one considers a supersymmetric theory [45–49], but we do not consider this an issue in our non-supersymmetric scenario.

We have made the simplifying assumption of a uniform isotropic universe and as such have ignored the spatial dependence of the scalar fields Φ and ϕ, and consequently that of the phase θ, in our analysis. In reality, as the universe expands, different parts of the universe will lose causal contact with each other, possibly lead-ing to the evolutions of Φ and θ obtaining spatial dependencies. More work needs to be done to see how much of an impact, if any, this can have on the baryon number density generated.

More analysis needs to be done on the possible chaotic nature of this mechanism, given its parallels with the forced pendulum. Preliminary work into the Poincare maps associated with the nature of this scenario have been conducted, and will be discussed in future work. Also, a further refinement of the numerical procedure needs to be considered to see whether lower reheating temperatures, corresponding to longer integration times, can be calculable.

Hopefully, through further investigation of this mechanism we can get a better understanding of the allowed regions of the parameter space. We also plan to explore the application of the ratchet mechanism in other cosmological settings, for example during the radiation epoch following the end of reheating.

References

1. K. Bamba, N.D. Barrie, A. Sugamoto, T. Takeuchi, K. Yamashita, Ratchet baryogenesis with an analogy to the forced pendulum (2016), arXiv:1610.03268
2. A. Albrecht, P.J. Steinhardt, M.S. Turner, F. Wilczek, Reheating an inflationary universe. Phys. Rev. Lett. **48**, 1437 (1982). https://doi.org/10.1103/PhysRevLett.48.1437
3. L. Kofman, A.D. Linde, A.A. Starobinsky, Reheating after inflation. Phys. Rev. Lett. **73**, 3195–3198 (1994). https://doi.org/10.1103/PhysRevLett.73.3195
4. Y. Shtanov, J.H. Traschen, R.H. Brandenberger, Universe reheating after inflation. Phys. Rev. D **51**, 5438–5455 (1995). https://doi.org/10.1103/PhysRevD.51.5438
5. A. Dolgov, K. Freese, Calculation of particle production by Nambu Goldstone bosons with application to inflation, reheating and baryogenesis. Phys. Rev. D **51**, 2693–2702 (1995). https://doi.org/10.1103/PhysRevD.51.2693
6. L.A. Kofman, The Origin of matter in the universe: reheating after inflation (1996), arXiv:9605155
7. L. Kofman, A.D. Linde, A.A. Starobinsky, Towards the theory of reheating after inflation. Phys. Rev. D **56**, 3258–3295 (1997). https://doi.org/10.1103/PhysRevD.56.3258
8. D.J.H. Chung, E.W. Kolb, A. Riotto, Production of massive particles during reheating. Phys. Rev. D **60**, 063504 (1999). https://doi.org/10.1103/PhysRevD.60.063504
9. S. Davidson, S. Sarkar, Thermalization after inflation. JHEP **11**, 012 (2000). https://doi.org/10.1088/1126-6708/2000/11/012
10. E.W. Kolb, A. Notari, A. Riotto, On the reheating stage after inflation. Phys. Rev. D **68**, 123505 (2003). https://doi.org/10.1103/PhysRevD.68.123505
11. B.A. Bassett, S. Tsujikawa, D. Wands, Inflation dynamics and reheating. Rev. Mod. Phys. **78**, 537–589 (2006). https://doi.org/10.1103/RevModPhys.78.537
12. R. Allahverdi, R. Brandenberger, F.-Y. Cyr-Racine, A. Mazumdar, Reheating in inflationary cosmology: theory and applications. Ann. Rev. Nucl. Part. Sci. **60**, 27–51 (2010). https://doi.org/10.1146/annurev.nucl.012809.104511
13. F. Takayama, Extremely long-lived charged massive particles as a probe for reheating of the universe. Phys. Rev. D **77**, 116003 (2008). https://doi.org/10.1103/PhysRevD.77.116003
14. J. Martin, C. Ringeval, First CMB constraints on the inflationary reheating temperature. Phys. Rev. D **82**, 023511 (2010). https://doi.org/10.1103/PhysRevD.82.023511
15. L. Dai, M. Kamionkowski, J. Wang, Reheating constraints to inflationary models. Phys. Rev. Lett. **113**, 041302 (2014). https://doi.org/10.1103/PhysRevLett.113.041302

16. G.F. Giudice, E.W. Kolb, A. Riotto, Largest temperature of the radiation era and its cosmological implications. Phys. Rev. D **64**, 023508 (2001). https://doi.org/10.1103/PhysRevD.64.023508

17. A. Dolgov, K. Freese, R. Rangarajan, M. Srednicki, Baryogenesis during reheating in natural inflation and comments on spontaneous baryogenesis. Phys. Rev. D **56**, 6155–6165 (1997). https://doi.org/10.1103/PhysRevD.56.6155

18. E.W. Kolb, M.S. Turner, The early universe. Front. Phys. **69**, 1–547 (1990)

19. A.A. Starobinsky, A new type of isotropic cosmological models without singularity. Phys. Lett. B **91**, 99–102 (1980). https://doi.org/10.1016/0370-2693(80)90670-X

20. V.F. Mukhanov, G.V. Chibisov, Quantum fluctuations and a nonsingular universe. JETP Lett. **33**, 532–535 (1981). [Pisma Zh. Eksp. Teor. Fiz. 33, 549 (1981)]

21. G. Magnano, M. Ferraris, M. Francaviglia, Nonlinear gravitational Lagrangians. Gen. Relativ. Gravit. **19**, 465 (1987). https://doi.org/10.1007/BF00760651

22. P.A.R. Ade et al., Planck 2015 results. XX. Constraints on inflation. Astron. Astrophys. **594**, A20 (2016b). https://doi.org/10.1051/0004-6361/201525898

23. D.S. Gorbunov, A.A. Tokareva, Inflation and reheating in the Starobinsky model with conformal Higgs Field. Phys. Part. Nucl. Lett. **10**, 633–636 (2013). https://doi.org/10.1134/S1547477113070030

24. J. Ellis, M.A.G. Garcia, D.V. Nanopoulos, K.A. Olive, Calculations of inflaton decays and reheating: with applications to no-scale inflation models. JCAP **1507**(07), 050 (2015). https://doi.org/10.1088/1475-7516/2015/07/050

25. S. Dimopoulos, L. Susskind, On the baryon number of the universe. Phys. Rev. D **18**, 4500–4509 (1978). https://doi.org/10.1103/PhysRevD.18.4500

26. I. Affleck, M. Dine, A new mechanism for baryogenesis. Nucl. Phys. B **249**, 361 (1985). https://doi.org/10.1016/0550-3213(85)90021-5

27. A.G. Cohen, D.B. Kaplan, Spontaneous baryogenesis. Nucl. Phys. B **308**, 913–928 (1988). https://doi.org/10.1016/0550-3213(88)90134-4

28. M. Yoshimura, Towards resolution of hierarchy problems in a cosmological context. Phys. Lett. B **608**, 183–188 (2005). https://doi.org/10.1016/j.physletb.2005.01.025

29. K. Bamba, M. Yoshimura, Curvaton scenario in the presence of two dilatons coupled to the scalar curvature. Prog. Theor. Phys. **115**, 269–308 (2006). https://doi.org/10.1143/PTP.115.269

30. A. Minamizaki, A. Sugamoto, Can the baryon number density and the cosmological constant be interrelated? Phys. Lett. B **659**, 656–660 (2008). https://doi.org/10.1016/j.physletb.2007.11.052

31. P. Reimann, R. Bartussek, R. Häußler, P. Hänggi, Brownian motors driven by temperature oscillations. Phys. Lett. A **215**, 26–31 (1996). https://doi.org/10.1016/0375-9601(96)00222-8

32. T. Takeuchi, A. Minamizaki, A. Sugamoto, Ratchet model of baryogenesis, in *Strong Coupling Gauge Theories in LHC Era. Proceedings, International Workshop, SCGT 09*, Nagoya, Japan, 8–11 Dec 2009 (2011), pp. 378–384. https://doi.org/10.1142/9789814329521_0041

33. A.G. Cohen, D.B. Kaplan, Thermodynamic generation of the baryon asymmetry. Phys. Lett. B **199**, 251–258 (1987). https://doi.org/10.1016/0370-2693(87)91369-4

34. N.F. Pedersen, O.H. Soerensen, B. Dueholm, J. Mygind, Half-harmonic parametric oscillations in josephson junctions. J. Low Temp. Phys. **38**, 1–23 (1980). https://doi.org/10.1007/BF00115266

35. D. D'Humieres, M.R. Beasley, B.A. Huberman, A. Libchaber, Chatoic states and routes to chaos int he forced pendulum. Phys. Rev. A **26**, 3483–3496 (1982). https://doi.org/10.1103/PhysRevA.26.3483

36. S.Y. Khlebnikov, M.E. Shaposhnikov, The statistical theory of anomalous fermion number nonconservation. Nucl. Phys. B **308**, 885–912 (1988). https://doi.org/10.1016/0550-3213(88)90133-2

37. J.A. Harvey, M.S. Turner, Cosmological baryon and lepton number in the presence of electroweak fermion number violation. Phys. Rev. D **42**, 3344–3349 (1990). https://doi.org/10.1103/PhysRevD.42.3344

38. A.G. Cohen, D.B. Kaplan, A.E. Nelson, Progress in electroweak baryogenesis. Ann. Rev. Nucl. Part. Sci. **43**, 27–70 (1993). https://doi.org/10.1146/annurev.ns.43.120193.000331
39. T. Yanagida, Horizontal symmetry and masses of neutrinos. Conf. Proc. C **7902131**, 95–99 (1979)
40. T. Yanagida, Horizontal symmetry and masses of neutrinos. Prog. Theor. Phys. **64**, 1103 (1980). https://doi.org/10.1143/PTP.64.1103
41. P. Ramond, The family group in grand unified theories, in *International Symposium on Fundamentals of Quantum Theory and Quantum Field Theory Palm Coast*, Florida, February 25–March 2, 1979, pp. 265–280
42. V.A. Kuzmin, V.A. Rubakov, M.E. Shaposhnikov, On the anomalous electroweak baryon number nonconservation in the early universe. Phys. Lett. B **155**, 36 (1985). https://doi.org/10.1016/0370-2693(85)91028-7
43. M. Trodden, Electroweak baryogenesis. Rev. Mod. Phys. **71**, 1463–1500 (1999). https://doi.org/10.1103/RevModPhys.71.1463
44. M. Fukugita, T. Yanagida, Baryogenesis without grand unification. Phys. Lett. B **174**, 45–47 (1986). https://doi.org/10.1016/0370-2693(86)91126-3
45. T. Moroi, H. Murayama, M. Yamaguchi, Cosmological constraints on the light stable gravitino. Phys. Lett. B **303**, 289–294 (1993). https://doi.org/10.1016/0370-2693(93)91434-O
46. M. Bolz, A. Brandenburg, W. Buchmuller, Thermal production of gravitinos. Nucl. Phys. B **606**, 518–544 (2001). https://doi.org/10.1016/S0550-3213(01)00132-8, https://doi.org/10.1016/j.nuclphysb.2007.09.020. [Erratum: Nucl. Phys. B 790, 336 (2008)]
47. M. Kawasaki, K. Kohri, T. Moroi, Big-Bang nucleosynthesis and hadronic decay of long-lived massive particles. Phys. Rev. D **71**, 083502 (2005). https://doi.org/10.1103/PhysRevD.71.083502
48. M. Viel, J. Lesgourgues, M.G. Haehnelt, S. Matarrese, A. Riotto, Constraining warm dark matter candidates including sterile neutrinos and light gravitinos with WMAP and the Lyman-alpha forest. Phys. Rev. D **71**, 063534 (2005). https://doi.org/10.1103/PhysRevD.71.063534
49. F. Takahashi, M. Yamada, Spontaneous baryogenesis from asymmetric inflaton. Phys. Lett. B **756**, 216–220 (2016). https://doi.org/10.1016/j.physletb.2016.03.020

Chapter 5
Gravitational Waves and the Cosmic Neutrino Background

The Cosmic Neutrino Background (CνB) contains information from very early times which may help illuminate both the properties of the neutrino sector and the evolution of the universe. Unfortunately, the weakly interacting nature of neutrinos combined with the low temperature of the background today, makes the prospect for detection near impossible in the foreseeable future. Despite this, the dynamics of the CνB could have had significant effects on the evolution of the early universe. The prospect of gleaning indirect evidence of the CνB is to be explored in this chapter, through considering the possible implications for gravitational wave propagation. Given the dawn of the new era of gravitational wave astronomy, this is an exciting possibility.

We argue that a CνB that carries a non-zero lepton charge develops gravitational instabilities, which are fundamentally related to the mixed gravity-lepton number anomaly. In the presence of this background, we find that a gravitational Chern-Simons (CS) term is induced, which leads to interesting physical effects. Firstly, gravitational waves propagating in such a neutrino background exhibit birefringent behaviour leading to an enhancement or suppression of the gravitational wave amplitudes, depending on the polarisation, with the magnitude of this effect related to the size of the lepton asymmetry. Secondly, this modification can lead to negative energy graviton modes in the high frequency regime, which induce very fast vacuum decays that produce, for example, positive energy photons and negative energy gravitons. Both of these effects can provide bounds on the lepton asymmetry of the universe, and hence probe the dynamics of the early universe [1].

5.1 The Cosmic Neutrino Background

Along with the CMB, the existence of the CνB is an inescapable prediction of the standard hot big bang cosmology [2]. At early times in the universe, the neutrinos are in thermal equilibrium with the SM plasma through the weak interactions alone.

© Springer International Publishing AG, part of Springer Nature 2018
N. D. Barrie, *Cosmological Implications of Quantum Anomalies*,
Springer Theses, https://doi.org/10.1007/978-3-319-94715-0_5

As the cross sections for these interactions are small compared to electromagnetic processes, the neutrinos will fall out of equilibrium well before the first generation of charged SM species. As given in Table 1.3, this occurs just prior to BBN, at around 2–3 MeV—with there being a temperature range due to the different decoupling times for each neutrino species. This is due to the additional weak interactions between electron neutrinos and the electrons and positrons present in the primordial plasma, which are not present for the muon and tau neutrinos because at these temperatures the population of charged muons and taus is thermally suppressed. The CνB is assumed to be a highly homogeneous and isotropic distribution of relic neutrinos with the characteristic temperature,

$$T_\nu = \left(\frac{4}{11} \right)^{\frac{1}{3}} T_\gamma \approx 1.945 \, \text{K} , \qquad (5.1)$$

where $T_\gamma = 2.725$ K is the temperature of the CMB today. Unlike the CMB though, the CνB is extremely hard to detect and its properties are largely unknown. The reason for the difference between the temperatures of the CνB and CMB, despite both sectors evolving as radiation for most of the cosmological history, is the extra entropy injection produced by electron-positron annihilation dominantly into photons after neutrino decoupling.

The physics of the generation and evolution of the CνB has been a closely studied area due to the window it could provide to early universe physics [3–5]. One interesting characteristic of the CνB is that it may exhibit a neutrino-antineutrino asymmetry. Unlike the baryon asymmetry which is strongly constrained, the CνB asymmetry can be relatively large. The associated lepton asymmetry is defined as follows,

$$\eta_{\nu_\alpha} = \frac{n_{\nu_\alpha} - \bar{n}_{\nu_\alpha}}{n_\gamma} \simeq \frac{\pi^2}{12\zeta(3)} \left(\xi_\alpha + \frac{\xi_\alpha^3}{\pi^2} \right) , \qquad (5.2)$$

for each neutrino flavour $\alpha = e, \mu, \tau$. Here $\xi_\alpha = \mu_\alpha / T$ is the degeneracy parameter, μ_α being the chemical potential for α-neutrinos. In fact, such an asymmetry is generically expected to be of the order of the observed baryon-antibaryon asymmetry, $\eta_B = (n_B - \bar{n}_B)/n_\gamma \sim 10^{-10}$, due to the equilibration by sphalerons of the lepton and baryon asymmetries in the very early universe, which conserve $B - L$. However, there are also models [6, 7] which predict an asymmetry in the neutrino sector that is many orders of magnitude larger than η_B. If so, this would have interesting cosmological implications for the QCD phase transition [8] and large-scale magnetic fields [9]. There have been attempts to circumvent the sphaleron redistribution constraints in such scenarios [6], which include the possible suppression of the equilibrium sphaleron processes when the lepton asymmetry is very large, $\xi \gg 1$ [10]. The presence of a large asymmetry can alter the cosmological expansion history, as it changes the effective number of relativistic species N_{eff}, due to the increased energy density in the neutrino sector associated with a non-zero neutrino chemical potential. That is,

$$\Delta N_{\text{eff}} \simeq \frac{30}{7} \left(\frac{\xi_\nu}{\pi} \right)^2 + \frac{15}{7} \left(\frac{\xi_\nu}{\pi} \right)^4 , \qquad (5.3)$$

which can be constrained by measurements of the CMB.

The most stringent bound on the relic lepton asymmetry comes from the successful theory of BBN. The BBN observables primarily constrain the electron neutrino asymmetry, due to the implications of a large electron neutrino asymmetry on the helium abundance [2, 11–24]. However, this bound applies to all flavours, since neutrino oscillations below $\sim$10 MeV are sizeable enough to lead to an approximate flavour equilibrium before BBN, $\mu_e \approx \mu_\mu \approx \mu_\tau (\equiv \mu_\nu)$ [25–27]. Although it has been found in a recent analysis that larger η_{ν_μ, ν_τ} asymmetries may be allowed [28]. In any case, the updated analysis in [22] leads to the following bound on the common degeneracy parameter,

$$|\xi_\nu| \lesssim 0.049 . \qquad (5.4)$$

The lack of strong constraints on the size of the relic lepton asymmetry has led to the postulation of many ideas associated with the prospect of having a large asymmetry, and mechanisms that can generate it, while not being in conflict with BBN constraints [28]. If it would become possible to observe the size of the asymmetry this would then be a potential smoking gun for these unorthodox neutrino physics models [7, 29–42].

The experimental observation of neutrino oscillations demonstrates that the neutrinos must carry mass, but the mechanism for the origin of these masses is not explained within the SM, and is still unknown. An interesting possibility is that the SM neutrinos have Majorana mass terms,

$$\mathcal{L}_m = \frac{1}{2} m_\nu \bar{\nu}_L^c \nu_L , \qquad (5.5)$$

where no right-handed neutrinos are present. In this scenario, the SM exhibits a mixed gravity-lepton number quantum anomaly.

If the neutrinos have Majorana masses, as we assume in this chapter, then the cosmological leptonic asymmetry carried by the CνB is defined as the difference between the left-handed neutrinos and right-handed antineutrinos [30, 43]. Due to this Majorana mass term, which violates lepton number, the relic neutrino asymmetry would be expected to be washed out once the mixing process becomes important. This would occur when the mixing timescale becomes smaller than that associated with the spatial expansion $\frac{1}{H}$. This means that at very early times a large Majorana neutrino asymmetry could be present and consistent with current observations, assuming that the rate of lepton number violation is such that it is sufficiently washed out before the freeze-out of equilibrium sphaleron processes.

5.2 Gravitational Waves and the Graviton Propagator

The recent observation of gravitational waves from black hole merger events by the LIGO collaboration [44], signals the beginning of the new era of gravitational wave astronomy [45, 46]. An important feature of upcoming gravitational wave astronomy experiments will be the ability to differentiate polarisations, hence allowing the exploration of possible astrophysical and cosmological birefringent effects. These can be the results or signs of very interesting physics [47–50], including CS modified gravity theories [51, 52]. Future gravitational wave detectors, such as eLISA [53, 54], will be able to probe the polarisations of incoming signals from astrophysical sources.

5.2.1 Linearised Gravitational Waves

In order to obtain predictions associated with gravitational wave production and propagation we utilise the linearised gravity approach. Gravitational waves can be described as small perturbations around a general spacetime background. The assumption of a small amplitude allows the higher order terms in h, the metric perturbations, to be neglected, removing non-linear interaction terms—producing the linearised gravity description. The metric perturbations are defined as follows,

$$g_{\mu\nu} \simeq \bar{g}_{\mu\nu} + \frac{1}{M_p} h_{\mu\nu} , \qquad (5.6)$$

where $h_{\mu\nu}$ is the metric perturbation and satisfies $|h_{\mu\nu}| \ll M_p$, and $M_p = 1/\sqrt{8\pi G}$ is the reduced Planck mass. In the linearised approximation, the Einstein-Hilbert action becomes,

$$
\begin{aligned}
S_\Lambda &= \int dx^4 \sqrt{-g} \, \frac{M_p^2}{2} R \\
&= \int dx^4 \sqrt{-g} \left(\frac{1}{2} h_{\mu\nu} \Box h^{\mu\nu} - h_{\mu\nu} \partial^\mu \partial_\alpha h^{\nu\alpha} + h \partial_\mu \partial_\nu h^{\mu\nu} - \frac{1}{2} h \Box h \right) \quad (5.7) \\
&= \frac{1}{2} \int dx^4 \sqrt{-g} \, \bar{h}_{\mu\nu} \Box \bar{h}^{\mu\nu} , \qquad (5.8)
\end{aligned}
$$

where in Eq. (5.8) we have taken $\bar{h}_{\mu\nu} = h_{\mu\nu} - \frac{1}{2} h \eta_{\mu\nu}$ and assumed the harmonic gauge, $\partial_\mu \bar{h}^\mu_\nu = \frac{1}{2} \partial_\nu \bar{h}$. The graviton coupling to matter is of the form $\mathcal{L}_{\text{int}} \propto h^{\mu\nu} T_{\mu\nu}$.

From this action we can derive the propagator for the graviton. In this chapter, we wish to consider the alteration to the graviton propagator in the presence of an asymmetric CνB. We shall do this by calculating the graviton polarisation tensor, which is connected to the graviton action in the following way,

$$S_\Lambda = \frac{1}{2} \int dx^4 \sqrt{-g} \, \bar{h}^{\mu\nu} \Pi_{\mu\nu\rho\sigma} \bar{h}^{\rho\sigma} , \tag{5.9}$$

where $\Pi_{\mu\nu\rho\sigma}$ is the graviton polarisation tensor.

The graviton h can be split into circularly polarised planewaves—h_R and h_L—in an isotropic and homogeneous universe, similarly to the quantised $U(1)_X$ gauge boson in Chap. 3. This shall be useful for our analysis of birefringent effects in this chapter.

Gravitational waves can have many sources, with both present-day and primordial origins being of interest. The recent measurement of black hole mergers is an example of present-day sources that correspond to very high energy scenarios. Cosmological sources may include phase transitions in the early universe, which can produce a stochastic gravitational wave background. Also, inflation generates tensor modes that can be imprinted in the B-mode polarisation of the CMB photons.

5.3 Graviton Polarisation Tensor in a Lepton Asymmetric CνB

In what follows we wish to consider the possible observational implications of a lepton asymmetric CνB on the properties of gravitational waves, and the possible inducement of gravitational instabilities. A non-zero lepton asymmetry for active neutrinos implies an imbalance between neutrinos of left-handed chirality and antineutrinos of right-handed chirality, and as we shall demonstrate, leads to the inducement of the gravitational CS term in the effective gravitational action. The possible effects of a large lepton asymmetry on primordial gravitational radiation has been considered in other contexts [55], and gravity has also been utilised to produce a relic neutrino asymmetry [56].

To determine the implications of a universal lepton asymmetry on gravity, we want to consider the effects on the graviton propagator. To do this we shall consider the presence of a chiral chemical potential μ_ν, which parametrises this asymmetry, and the gravity-lepton number chiral quantum anomaly. This anomaly is present in the SM when considering Majorana neutrinos; due to the absence of right-handed neutrinos. In this scenario we can expect an induced parity violating contribution to the gravitational action, which can be found through calculating the contribution of the chiral chemical potential to the graviton polarisation tensor.

We calculate the inducement of the CS like term to the effective graviton Lagrangian through the 1-loop graviton polarization diagram depicted in Fig. 5.1, which is influenced by the chemical potential μ_ν. The lepton asymmetry is enforced in the Lagrangian through the chiral chemical potential by the following term, $\mathcal{L}_{\mu_\nu} = \bar{\nu}\slashed{b}\gamma^5\nu = \mu_\nu\bar{\nu}\gamma_0\gamma^5\nu$, where we have considered the frame in which the CνB is at rest ($\slashed{b} = \mu_\nu\gamma_0$). The neutrino propagator is altered as follows,

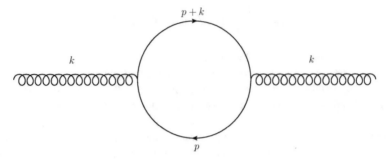

Fig. 5.1 The 1-loop correction to the graviton propagator that is relevant in the parity violating background, with chemical potential μ_ν

$$S(p) = \frac{i}{\not{p} - m - \not{b}\gamma^5} = \frac{i}{\not{p} - m} \sum_{n=0}^{\infty} \left(-i\not{b}\gamma^5 \frac{i}{\not{p} - m}\right)^n \equiv S_0(p) + \sum_{n=1}^{\infty} S_n(p),$$
(5.10)

where $S_0(p)$ is the usual fermion propagator in vacuum.

In the calculation that follows, we shall use the standard Feynman rules [57, 58], with the above modified neutrino propagator to first-order in μ_ν, $S(p) \approx S_0(p) - i\mu_\nu \frac{i}{\not{p}-m}\gamma_0\gamma^5 \frac{i}{\not{p}-m}$. The higher order terms in b_μ, or μ_ν, are neglected because we are only interested in the linear terms in b_μ, which give the CS like term. Thus we find that the induced parity odd part of the graviton polarization tensor is,

$$\Pi_{\mu\nu\rho\sigma} = -\int \frac{d^4 p}{(2\pi)^4} (2p + k)_\nu (2p + k)_\sigma \left[Tr(\gamma_\mu S_0(p+k)\gamma_\rho S_1(p)) \right. \tag{5.11}$$

$$\left. + Tr(\gamma_\rho S_0(p)\gamma_\mu S_1(p+k)) \right]. \tag{5.12}$$

To evaluate the divergent loop integral in Eq. (5.12) we employ the dimensional regularization method ($d = 4 - \epsilon, \epsilon \to 0$) and use the relations provided in Appendix C. This gives,

$$\Pi_{\mu\nu\rho\sigma} = \frac{\mu_\nu}{8\pi^2} k^\alpha \varepsilon_{\mu\rho\alpha0} \int_0^1 dx \left[\frac{4\pi^2\lambda^2}{M^2}\right]^\epsilon \left[8x^2(1-x)^2(1-2x)^2 \frac{k^2}{M^2}\Gamma(1+\epsilon)k_\nu k_\sigma \right.$$
$$+ (24x^2 - 44x + 18)\Gamma(\epsilon-1)M^2\eta_{\nu\sigma} - 16x^2(1-x)^2\Gamma(\epsilon)k^2\eta_{\nu\sigma}$$
$$\left. - (80x^4 - 192x^3 + 156x^2 - 50x + 5)\Gamma(\epsilon)k_\nu k_\sigma \right], \tag{5.13}$$

where $M^2 = m^2 - x(1-x)k^2$ and the limit $\epsilon \to 0$ is assumed. In simplifying this result we find a divergent quantity that is of the following form,

$$\Pi_{\mu\nu\rho\sigma}^{(div)} = -\frac{1}{\epsilon}\frac{\mu_\nu}{2\pi^2}k^\alpha \varepsilon_{\mu\rho\alpha0}m^2\eta_{\nu\sigma}, \tag{5.14}$$

where γ is Euler's constant. A straightforward inspection reveals that this divergent term does not satisfy the gravitational Ward identity, $k^\nu \Pi^{(div)}_{\mu\nu\rho\sigma} \neq 0$, and hence violates the gauge invariance of the effective gravitational action. This has also been observed previously in related calculations [59, 60]. The origin of this violation is rooted in the method of dimensional regularization, which violates local Lorentz invariance explicitly through the extrapolation to a non-integer number of spacetime dimensions $d = 4 - \epsilon$. Therefore, following the standard lore, we introduce non-invariant counter-terms to renormalise away this unphysical divergent term. The polarisation tensor then takes the following simple form,

$$\Pi_{\mu\nu\rho\sigma} = \mu_\nu \varepsilon_{\mu\rho\alpha 0} k^\alpha [k_\nu k_\sigma - k^2 \eta_{\nu\sigma}] C(k^2) , \tag{5.15}$$

where

$$C(k^2) = \frac{1}{192\pi^2} - \frac{m^2}{16\pi^2 (k^2)^{3/2}} \left[\sqrt{k^2} - \sqrt{4m^2 - k^2} \tan^{-1}\left(\frac{\sqrt{k^2}}{\sqrt{4m^2 - k^2}}\right) \right] . \tag{5.16}$$

This further reduces to,

$$C(k^2) = \begin{cases} -\frac{1}{1920\pi^2} \frac{k^2}{m^2}, & \text{if } k^2/m^2 \ll 1 \\ \frac{1}{192\pi^2}, & \text{if } k^2/m^2 \gg 1 \end{cases} . \tag{5.17}$$

From this polarization tensor we can determine the induced parity violating term in the effective action. We shall now investigate the two possible limiting cases separately.

In the limit $k^2/m_\nu^2 \ll 1$, this term reads,

$$S_{eff} \propto \frac{\mu_\nu}{m_\nu^2} \int d^4 x \varepsilon_{\mu\rho\alpha 0} h^{\mu\nu} \partial^\alpha \Box (\Box h^{\rho\sigma} \eta_{\nu\sigma} - \partial_\nu \partial_\sigma h^{\rho\sigma}) . \tag{5.18}$$

In the case of the harmonic gauge, taking $\bar{h}^{\mu\nu} = h^{\mu\nu} - \frac{1}{2}\eta^{\mu\nu} h$, for which $\partial_\mu \bar{h}^{\mu\nu} = 0$, the action reduces to,

$$S \propto \frac{\mu_\nu}{m_\nu^2} \int d^4 x \varepsilon_{\mu\rho\alpha 0} \bar{h}^{\mu\nu} \partial^\alpha \Box^2 \bar{h}^{\rho\sigma} \eta_{\nu\sigma} . \tag{5.19}$$

Note that this induced term contains more than two derivatives and thus is significant only in the ultraviolet regime. Indeed, taking the harmonic gauge we derive the modified equation of motion for the linearised graviton field,

$$\Box \bar{h}_{ij} = \frac{1}{1920\pi^2} \epsilon_{ilk} \partial^l \frac{\mu_\nu}{m_\nu^2 M_p^2} \Box^2 \bar{h}^k_{\ j} . \tag{5.20}$$

The dispersion relations for left $(-)$ and right $(+)$ graviton polarisation modes then readily follow from Eq. (5.20),

$$(\omega^2 - |\mathbf{k}|^2) \mp \frac{\mu_\nu}{1920\pi^2 m_\nu^2 M_p^2} |\mathbf{k}|(\omega^2 - |\mathbf{k}|^2)^2 = 0 . \tag{5.21}$$

One can then see that for large enough momenta $|\mathbf{k}|$ there are modes with imaginary frequencies,

$$(\omega^2 - |\mathbf{k}|^2) = -\frac{1920\pi^2 m_\nu^2 M_p^2}{\mu_\nu |\mathbf{k}|} , \tag{5.22}$$

which are unstable. Such potentially unstable modes, however, have extremely small wavelengths,

$$|\mathbf{k}| \gtrsim \frac{1920\pi^2 M_p^2}{\mu_\nu} . \tag{5.23}$$

In this trans-Planckian regime, the Einstein's theory itself is believed to be untrustworthy, so we do not consider this limit any further.

It is more interesting to consider the opposite limit, $k^2/m_\nu^2 \gg 1$. In this limit we obtain the following contribution to the graviton action,

$$\begin{aligned} S_{eff} &= -\frac{\mu_\nu}{192\pi^2} \int d^4x \varepsilon_{\mu\rho\alpha 0} h^{\mu\nu} \partial^\alpha (\Box h^{\rho\sigma} \eta_{\nu\sigma} - \partial_\nu \partial_\sigma h^{\rho\sigma}) \\ &= \frac{\mu_\nu}{48\pi^2} \int d^4x \, K^0 , \end{aligned} \tag{5.24}$$

which contains the same number of derivatives as the standard kinetic term in the weak field approximation. In fact, K^0 is the linearised 0th component of the four dimensional CS topological current,

$$K^\beta = \varepsilon^{\beta\alpha\mu\nu} (\Gamma^\sigma_{\alpha\rho} \partial_\mu \Gamma^\rho_{\nu\sigma} - \frac{2}{3} \Gamma^\sigma_{\alpha\rho} \Gamma^\rho_{\mu\lambda} \Gamma^\lambda_{\nu\sigma}) . \tag{5.25}$$

Therefore, the presence of an asymmetry in the CνB replicates CS modified gravity,

$$S_{CS} = \int d^4x \, (\partial_\mu \theta) K^\mu = \int d^4x \, \theta (^*RR) , \tag{5.26}$$

where we can make the identification $\partial_0 \theta = \frac{\mu_\nu}{48\pi^2}$ and,

$$^*RR := R\tilde{R} = {^*R^\mu}_\nu{}^{\rho\sigma} R^\nu{}_{\mu\rho\sigma} , \tag{5.27}$$

where the dual tensor is given by ${^*R^\mu}_\nu{}^{\rho\sigma} := \frac{1}{2}\varepsilon^{\rho\sigma\alpha\beta} R^\nu{}_{\mu\alpha\beta}$.

This is the expected result in a parity violating background, and is found in both gravitational and electromagnetic cases [58–69].

5.4 Chern-Simons Modified Gravity and Observational Implications

CS modified gravity is an area of active research due to it naturally appearing in the context of anomaly cancellation in String Theory, via the Green-Schwarz mechanism [70], and in Loop Quantum Gravity, as well as providing many interesting phenomenological implications [51, 52]. It is also a useful model independent way of parametrising parity violating effects in the cosmological setting. This modification to the gravitational action is analogous to the CS terms associated with the Green-Schwarz mechanism utilised in Chap. 3, but with the Riemann tensor replacing the field strength tensor. There are many potential observational consequences of this extension that have been explored, from which limits have been imposed on the CS coupling parameter.

The terms added to the gravitational action for CS modified gravity are given below,

$$S_{CS} \propto \int d^4x \sqrt{-g} \theta^* RR , \qquad (5.28)$$

where $\kappa = 1/8\pi G$, and θ denotes the CS coupling, which we have assumed to be associated with a non-dynamical CS extension. In the case of a dynamical CS extension, a kinetic and potential term for θ are required,

$$S_\theta \propto \int d^4x \sqrt{-g} [g^{\mu\nu} (\nabla\theta_\mu)(\nabla\theta_\nu) + 2V(\theta)] . \qquad (5.29)$$

Although, experimental tests of CS modified gravity have been mostly confined to the non-dynamical scenario [71–74]. This is due to the increased complexity associated with the dynamical case, which has not been explored as extensively in the literature. Fortunately, the case we consider in our work is of the non-dynamical type, and hence we can consider the effects that have been explored in this simpler scenario.

An interesting phenomenon associated with CS gravity is birefringent propagation of gravitational waves. The CS correction leads to an exponential enhancement and suppression of the left and right circularly polarised waves, which depend on the wave number, CS coupling and the integrated history of the propagation [52].

5.4.1 Birefringent Propagation Through an Asymmetric CνB

Planned gravitational wave detectors, such as eLISA, DECIGO and BBO, can potentially measure the polarization of observed gravitational waves, and hence potentially this birefringence effect. We now wish to consider the magnitude of the birefringent effect induced by the CS term in Eq. (5.24), and determine the possible sources that could be constrained by this effect. To this end, we parametrise the gravitational waves as, $h_{ij} = \frac{A_{ij}}{a(\eta)} \exp[-i(\phi(\eta) - \mathbf{k} \cdot \mathbf{x})]$, which can be decomposed into the two circularly polarised states—e_{ij}^R and e_{ij}^L. The two possible circularly polarised states are, $e_{ij}^R = \frac{1}{\sqrt{2}}(e_{ij}^+ + i e_{ij}^\times)$ and $e_{ij}^L = \frac{1}{\sqrt{2}}(e_{ij}^+ - i e_{ij}^\times)$, which satisfy $n_i \varepsilon^{ijk} e_{kl}^{R,L} = i \lambda_{R,L} (e_l^j)^{R,L}$, where $\lambda_{R,L} = \pm 1$. The phase factor $\lambda_{R,L}$ leads to exponential suppression or enhancement of the left and right circular polarisations of the the propagating gravitational waves, the magnitude of which we shall now calculate. From the equations of motion for the action $S = S_{EH} + S_{eff}$ we obtain, for a general θ [75],

$$(i\phi_{,\eta\eta}^{R,L} + (\phi_{,\eta}^{R,L})^2 + \mathcal{H}_{,\eta} + \mathcal{H}^2 - |\mathbf{k}|^2)\left(1 - \frac{\lambda_{R,L}\kappa\theta_{,\eta}}{a^2}\right)$$
$$= \frac{i\lambda_{R,L}|\mathbf{k}|}{a^2}(\theta_{,\eta\eta} - 2\mathcal{H}\theta_{,\eta})(\phi_{,\eta}^{R,L} - i\mathcal{H}). \qquad (5.30)$$

We first solve the above equation assuming propagation in the matter dominated epoch, $a(\eta) = a_0\eta^2 = \frac{a_0}{1+z}$, as we are considering sources within the range of possible near future observatories—$z < 30$. The accumulated phase over propagation, to first order in θ, is given by,

$$\Delta\phi_{mat}^{R,L} = i\lambda_{R,L}|\mathbf{k}|H_0 \int_\eta^1 \left[\frac{1}{4}\theta_{,\eta\eta} - \frac{1}{\eta}\theta_{,\eta}\right]\frac{d\eta}{\eta^4}. \qquad (5.31)$$

In the case considered in this work, we make the following identification $\theta_{,\eta} = \left(\frac{a(\eta_0)}{a(\eta)}\right)^2 \frac{\mu_0}{48\pi^2 M_p^2}$, where $\mu_0 = a(\eta)\mu_\nu$ is the chemical potential at present. Thus, for the asymmetric CνB,

$$\Delta\phi_{mat}^{R,L} = -i\frac{1}{288\pi^2}\frac{\mu_\nu H_0}{M_p^2}\left(\frac{|\mathbf{k}|}{1\text{ GeV}}\right)(1+z)^4. \qquad (5.32)$$

Hence the ratio of the wave amplitudes of the two polarisation states is given by,

$$\frac{h_R}{h_L} \propto e^{-2|\Delta\phi_{mat}^{R,L}|}. \qquad (5.33)$$

Taking into account the current bounds on the CνB asymmetry parameter, $\xi < 0.049$, we find $|i\Delta\phi^{R,L}| \lesssim 10^{-87}\left(\frac{|\mathbf{k}|}{1\text{ GeV}}\right)$, for $z \sim 30$. This accumulated phase difference is too small to be observable by any conceivable gravitational wave detector.

This leads us to shift to the more interesting scenario, which is the propagation of gravitational waves from sources in the very early universe. The effect would be thought to be of a higher magnitude because both the chemical potential would have been larger and the waves would have a longer propagation time over which a phase difference can be accumulated. Conceivably, any early universe sources could provide constraints, if the two polarisations of the signal are differentiable. Some examples of this would be the inflationary gravitational waves, and those produced by collisions of bubbles formed during phase transitions. Therefore, we now consider gravitational waves produced at very early times, during the radiation dominated epoch. The accumulated phase now reads,

$$\Delta\phi_{rad}^{R,L} = i\lambda_{R,L} \frac{|\mathbf{k}|}{\Omega_{r,0}H_0^2} \int_{\eta}^{1} \left[\frac{1}{2}\theta_{,\eta\eta} - \frac{1}{\eta}\theta_{,\eta}\right] \frac{d\eta}{\eta^2}, \tag{5.34}$$

where $\Omega_{r,0} \sim 9.2 \cdot 10^{-5}$ is the radiation density parameter today. After taking the integral we find,

$$\Delta\phi_{rad}^{R,L} \simeq -i\lambda_{R,L}\xi_\nu \left(\frac{|\mathbf{k}|}{1\,\text{GeV}}\right)\left(\frac{T_s}{1\,\text{TeV}}\right)^4, \tag{5.35}$$

where we have redefined the redshift in terms of the temperature at which the gravitational waves are produced, T_s, or when the asymmetry is generated, whichever is lowest.

From Eq. (5.35), it can be seen that if an asymmetry is present in the CνB, which equilibrium sphalerons transitions may assure, then it is possible to get significant birefringent behaviour in the propagation of gravitational waves from primordial sources, depending on the momenta $|\mathbf{k}|$ of the gravitational waves and size of the asymmetry.

If one is to assume that the characteristic momenta of the gravitational waves is of the order of the Hubble rate at the time of production ($|\mathbf{k}| \sim H$), and the asymmetry is already present, then we get the following interesting constraint on the gravitational waves produced from the source, at a temperature T_s,

$$\Delta\phi_{rad}^{R,L} \simeq -i\lambda_{R,L}\xi_\nu \left(\frac{T_s}{10^6\,\text{GeV}}\right)^5, \tag{5.36}$$

hence very early sources and a large asymmetry are required, if a relative enhancement or suppression of order one is to be observed.

5.5 Induced Ghost-Like Modes and Vacuum Decay

Another interesting consequence of the induced CS term in Eq. (5.24) is that short-scale gravitational fluctuations exhibit negative energy modes, which if present lead to a rapid decay of a vacuum state, for example, into negative energy graviton and photons [74]. Since in this setting the graviton energy would not be bounded from below, the phase space for this process is formally infinite [76, 77], and as such will develop very rapidly. We investigate the production of two photons and a negative energy graviton via this process, to provide constraints on the neutrino asymmetry at early times. The relevant effective interaction is of the form,

$$
\begin{aligned}
S_{\text{int}} &\sim \frac{1}{m_*} \int d^4 x h_{\mu\nu}^{can} T^{\mu\nu} \\
&= \frac{1}{m_*} \int d^4 x \frac{1}{2} h^{can} F_{\mu\nu} F^{\mu\nu} - h_{\mu\nu}^{can} F^{\mu\alpha} F_\alpha^\nu ,
\end{aligned}
\tag{5.37}
$$

where the canonically normalised graviton field is $h_{\mu\nu}^{can} = m_{can} h_{\mu\nu}^{can}$, with the definition,

$$
m_{can} = M_p \sqrt{1 + \lambda_{R,L} \frac{|\mathbf{k}|}{a m_{CS}}} ,
\tag{5.38}
$$

where m_{CS} is the analogous CS mass scale,

$$
m_{CS}(t) = \frac{M_p^2}{\mu_\nu} = \frac{M_p^2}{\xi T} = \frac{a(t) M_p^2}{\mu_0} .
\tag{5.39}
$$

5.5.1   Photon Energy Spectrum from Induced Vacuum Decay

To obtain a finite result for the decay rate we need to constrain the phase space. In the absence of a fundamental physical reason for such a truncation, we follow [76, 77], and simply cut-off the three momenta at $|\mathbf{k}|_{max} = \Lambda$. In the analysis that follows, we consider decays into this mode as it will have the largest contribution to the energy density of the generated photons. In addition, we take the reasonable approximation,

$$
m_{can} \simeq \sqrt{\frac{|\mathbf{k}| \mu_\nu}{a}} ,
\tag{5.40}
$$

and consider the dynamics of our scenario prior to BBN and after reheating, when the universe is radiation dominated and evolves as follows,

$$
a(t) = a_0 \sqrt{t} = \sqrt{2 \Omega_{r,0}^{1/2} H_0 t} ,
\tag{5.41}
$$

where $\Omega_{r,0} \sim 9.2 \cdot 10^{-5}$ is the radiation density parameter today.

   The time at which this ghost term is no longer present will be defined as t_* and is found in terms of the scale factor as,

$$
1 \simeq \frac{\Lambda}{a(t_*) m_{CS}(t_*)} \quad \Rightarrow \quad a(t_*) \simeq \sqrt{\frac{\mu_0 \Lambda}{M_p^2}} \text{ or } a(t_*) \simeq \frac{\xi_\nu T_* \Lambda}{M_p^2} ,
\tag{5.42}
$$

where T_* is the temperature at which the ghost terms stop contributing.

This fixes the time at which the ghost modes no longer exist, and decay of the vacuum ceases. We can reinterpret this as a temperature, so that it is possible to associate this with the maximal reheating temperature, and also ensure it does not have adverse implications on BBN. If we assume that the asymmetry is produced during or after the reheating epoch, and prior to BBN, the scale factor has a $\frac{1}{T}$ dependence if we ignore the decoupling of radiation degrees of freedom. The scale factor takes the following form,

$$a(t) \simeq \left(\frac{90\Omega_{r,0}}{g^*\pi^2}\right)^{\frac{1}{4}} \frac{\sqrt{H_0 M_p}}{T} \, , \tag{5.43}$$

where $g^* \simeq 106.75$. Equating Eqs. (5.42) and (5.43) to find the temperature at which this effect ends, we find,

$$T_* = \left(\sqrt{\frac{90\Omega_{r,0}}{g^*\pi^2}} \frac{H_0 M_p^3}{\xi_\nu^2}\right)^{\frac{1}{4}} \sqrt{\frac{M_p}{\Lambda}}$$

$$\simeq \frac{440}{\sqrt{\xi_\nu}} \, \text{GeV} \sqrt{\frac{M_p}{\Lambda}} \, . \tag{5.44}$$

Given that the maximum reheating temperature is $T_{\text{rh}} \sim 10^{15}$ GeV, Eq. (5.44) implies we can constrain the production temperature of neutrino asymmetries satisfying $\xi_\nu \gtrsim 2 \cdot 10^{-25} \frac{M_p}{\Lambda}$, with smaller ξ's not generating ghost like modes after reheating. We also assume here that ξ is approximately constant, and hence is the same parameter currently constrained by BBN measurements, in the calculation of the lepton asymmetry stored in the CνB.

Next we compute the spectrum of photons generated by the induced vacuum decay, and then subsequently the energy density, which can be constrained by experiment. It is given by,

$$\frac{1}{a^3} \frac{d}{dt}(a^3 n(k,t)) = \Gamma \delta\left(\frac{|\mathbf{k}|}{\Lambda} - 1\right) , \tag{5.45}$$

where $n(k,t)$ is the number of photons per unit logarithmic wave number $|\mathbf{k}|$ and Γ is the total decay width, which we take to approximately be,

$$\Gamma \sim \frac{\Lambda^6}{m_{can}^2} = \frac{a(t)\Lambda^6}{|\mathbf{k}|\mu_\nu} = \frac{a(t)^2\Lambda^5}{\mu_0} \, . \tag{5.46}$$

Since the above decay rate is much faster than the expansion rate of the universe, we may safely assume that the decay is approximately instantaneous. Therefore, we fix the scale factor in Eq. (5.46) at time t_a, when the asymmetric background is first produced. We then integrate Eq. (5.45) between the time t_a and when the ghost terms are no longer present, t_*,

$$|\mathbf{k}|n_*(|\mathbf{k}|) \sim \frac{a(t_*)^2 \Lambda \Gamma_a}{5\Omega_{r,0}^{1/2} H_0} . \tag{5.47}$$

Taking into account the dilution factor due to the expansion of the universe from the end of photon production to today, $\left(\frac{a(t_*)}{a_0}\right)^3 = a(t_*)^3$, we obtain,

$$|\mathbf{k}|n_0(|\mathbf{k}|) \sim \frac{a(t_*)^5 \Lambda \Gamma_a}{5\Omega_{r,0}^{1/2} H_0} . \tag{5.48}$$

Therefore, the energy density for a given momenta $|\mathbf{k}|$ is,

$$\frac{dE}{d^3x\, d\ln|\mathbf{k}|} \sim |\mathbf{k}|n_0(|\mathbf{k}|) \sim \frac{\xi^4 T_*^5}{10 T_a^2} \sqrt{\frac{M_p^3}{H_0}} \left(\frac{\Lambda}{M_p}\right)^{11} . \tag{5.49}$$

We can now obtain a conservative bound on the energy density in the produced photons, through the observation that the universe is not radiation dominated today,

$$\frac{dE}{d^3x\, d\ln|\mathbf{k}|} \lesssim M_p^2 H_0^2 . \tag{5.50}$$

From this we get the following constraint on ξ_ν, assuming the asymmetry is generated above the characteristic temperature T_*, when requiring consistency with observation,

$$\xi_\nu \lesssim 2 \cdot 10^{-41} \left(\frac{T_a}{10^{15}\,\text{GeV}}\right)^{4/3} \left(\frac{M_p}{\Lambda}\right)^{17/3} , \tag{5.51}$$

for which it is assumed $T_a \gtrsim \frac{440}{\sqrt{\xi_\nu}}\,\text{GeV}\sqrt{\frac{M_p}{\Lambda}}$. Equivalently,

$$T_* \gtrsim 10^{23}\,\text{GeV} \left(\frac{T_a}{10^{15}\,\text{GeV}}\right)^{-2/3} \left(\frac{\Lambda}{M_p}\right)^{17/6} . \tag{5.52}$$

Thus we arrive at the conclusion that, unless $\Lambda \ll M_p$, the resulting photon energy density from the induced vacuum decay can hardly be accommodated with observation. Substituting the constraint in Eq. (5.51) into that for the asymmetry stored in the CνB as a function of ξ_ν, in Eq. (5.2), we find the following bound,

$$\eta_\nu \lesssim 10^{-41} \left(\frac{T_a}{10^{15}\,\text{GeV}}\right)^{4/3} \left(\frac{M_p}{\Lambda}\right)^{17/3} . \tag{5.53}$$

If we instead assume that $T_a \lesssim \frac{440}{\sqrt{\xi_\nu}}\,\text{GeV}\sqrt{\frac{M_p}{\Lambda}}$, and hence vacuum decay does not occur, then we get the following constraint on η_ν,

$$\eta_\nu \lesssim 0.033 \left(\frac{2000\,\text{GeV}}{T_a} \right)^2 \frac{M_p}{\Lambda} , \tag{5.54}$$

where $\eta_\nu \lesssim 0.033$ is the current upper limit from BBN constraints.

5.6 Conclusions and Future Prospects

In this chapter, we have argued that a relic neutrino background with non-zero lepton number can lead to gravitationally observable effects [1]. We have explicitly calculated the parity odd part of the graviton polarization tensor in a lepton asymmetric CνB, which induces a gravitational CS term in the effective action. The observable implications of this were then explored, wherein the derived gravitational instabilities are related to the gravity-lepton number mixed anomaly.

The induced CS term leads to birefringent behaviour, causing an enhancement or suppression of the gravitational wave amplitudes depending on the polarisation. While this effect is negligible for local sources, we demonstrate that it could be sizeable for gravitational waves produced in the very early universe,

$$\Delta \phi_{rad}^{R,L} \simeq -i \lambda_{R,L} \xi_\nu \left(\frac{|\mathbf{k}|}{1\,\text{GeV}} \right) \left(\frac{T_s}{1\,\text{TeV}} \right)^4 , \tag{5.55}$$

which when considering a source produced with momenta $|\mathbf{k}| \sim H$, with H being the Hubble rate at the time of production,

$$\Delta \phi_{rad}^{R,L} \simeq -i \lambda_{R,L} \xi_\nu \left(\frac{T_s}{10^6\,\text{GeV}} \right)^5 , \tag{5.56}$$

which immediately indicates the need for very early sources, if interesting constraints are to be obtained.

In addition to the above, we have also argued that short-scale gravitational fluctuations in the presence of an asymmetric CνB exhibit negative energy modes, which can lead to the rapid decay of the vacuum state into negative energy graviton and positive energy photons. Since the graviton energy is not bounded from below, the phase space for this process is formally infinite, that is the instability is expected to develop very rapidly. Conservatively, we introduced a comoving cut-off Λ and computed the spectrum of produced photons as a function of the neutrino chemical potential. From the constraints on the radiation energy density today, we have obtained an interesting bound on the neutrino degeneracy parameter,

$$\xi_\nu \lesssim 2 \cdot 10^{-41} \left(\frac{T_a}{10^{15}\,\text{GeV}} \right)^{4/3} \left(\frac{M_p}{\Lambda} \right)^{17/3} , \tag{5.57}$$

which unless $\Lambda \ll M_p$, would effectively rule out the existence of an asymmetric CνB that is produced early enough for ghost modes to be present. If we assume there are no ghost modes associated with the CνB at any point in the early universe we obtain the following constraint,

$$\eta_\nu \lesssim 0.033 \left(\frac{2000\,\text{GeV}}{T_a} \right)^2 \frac{M_p}{\Lambda} \, . \qquad (5.58)$$

We believe that the findings reported in this chapter will prove to be useful for gaining a greater understanding of the properties of the CνB, and possibly allow constraints to be placed on particle physics models containing a lepton asymmetry. Being able to constrain or obtain indirect measurements of the size of the lepton asymmetry would help illuminate the properties of the neutrinos and possibly the mechanism for Baryogenesis.

This paper [1] explores the exciting new possibility of using gravitational wave phenomena to uncover information about the properties of the fundamental particles of nature. It will be interesting to consider these CνB effects further, and also if similar phenomena could be induced by dark matter or other potential relics.

References

1. N.D. Barrie, A. Kobakhidze, Gravitational instabilities of the cosmic neutrino background with non-zero lepton number. Phys. Lett. B **772**, 459–463 (2017). https://doi.org/10.1016/j.physletb.2017.07.012
2. J. Lesgourgues, S. Pastor, Cosmological implications of a relic neutrino asymmetry. Phys. Rev. D **60**, 103521 (1999). https://doi.org/10.1103/PhysRevD.60.103521
3. N.F. Bell, R.R. Volkas, Y.Y.Y. Wong, Relic neutrino asymmetry evolution from first principles. Phys. Rev. D **59**, 113001 (1999). https://doi.org/10.1103/PhysRevD.59.113001
4. R.R. Volkas, Y.Y.Y. Wong, Further studies on relic neutrino asymmetry generation. 1. The adiabatic Boltzmann limit, nonadiabatic evolution, and the classical harmonic oscillator analog of the quantum kinetic equations. Phys. Rev. D **62**, 093024 (2000). https://doi.org/10.1103/PhysRevD.62.093024
5. K.S.M. Lee, R.R. Volkas, Y.Y.Y. Wong, Further studies on relic neutrino asymmetry generation. 2. A Rigorous treatment of repopulation in the adiabatic limit. Phys. Rev. D **62**, 093025 (2000). https://doi.org/10.1103/PhysRevD.62.093025
6. J. March-Russell, H. Murayama, A. Riotto, The small observed baryon asymmetry from a large lepton asymmetry. JHEP **11**, 015 (1999). https://doi.org/10.1088/1126-6708/1999/11/015
7. J. McDonald, Naturally large cosmological neutrino asymmetries in the MSSM. Phys. Rev. Lett. **84**, 4798–4801 (2000). https://doi.org/10.1103/PhysRevLett.84.4798
8. D.J. Schwarz, M. Stuke, Lepton asymmetry and the cosmic QCD transition. JCAP **0911**, 025 (2009). https://doi.org/10.1088/1475-7516/2009/11/025,10.1088/1475-7516/2010/10/E01. [Erratum: JCAP1010, E01(2010)]
9. V.B. Semikoz, D.D. Sokoloff, J.W.F. Valle, Is the baryon asymmetry of the universe related to galactic magnetic fields? Phys. Rev. D **80**, 083510 (2009). https://doi.org/10.1103/PhysRevD.80.083510
10. G. Barenboim, W.I. Park, A full picture of large lepton number asymmetries of the universe. JCAP **1704**(04), 048 (2017). https://doi.org/10.1088/1475-7516/2017/04/048

11. A. Yahil, G. Beaudet, Big-bang nucleosynthesis with nonzero lepton numbers. Astrophys. J. **206**, 26–29 (1976). https://doi.org/10.1086/154352

12. N. Terasawa, K. Sato, Constraints on baryon and lepton number asymmetries of the early universe from primordial nucleosynthesis. Prog. Theor. Phys. **72**, 1262–1265 (1984). https://doi.org/10.1143/PTP.72.1262

13. N. Terasawa, K. Sato, Lepton and baryon number asymmetry of the universe and primordial nucleosynthesis. Prog. Theor. Phys. **80**, 468 (1988). https://doi.org/10.1143/PTP.80.468

14. K. Kohri, M. Kawasaki, K. Sato, Big bang nucleosynthesis and lepton number asymmetry in the universe. Astrophys. J. **490**, 72–75 (1997). https://doi.org/10.1086/512793

15. W.H. Kinney, A. Riotto, Measuring the cosmological lepton asymmetry through the CMB anisotropy. Phys. Rev. Lett. **83**, 3366–3369 (1999). https://doi.org/10.1103/PhysRevLett.83.3366

16. S. Pastor, J. Lesgourgues, Relic neutrino asymmetry, CMB and large scale structure. Nucl. Phys. Proc. Suppl. **81**, 47–51 (2000). https://doi.org/10.1016/S0920-5632(99)00857-9

17. J. Lesgourgues, S. Pastor, S. Prunet, Cosmological measurement of neutrino mass in the presence of leptonic asymmetry. Phys. Rev. D **62**, 023001 (2000). https://doi.org/10.1103/PhysRevD.62.023001

18. J. Lesgourgues, M. Peloso, Remarks on the Boomerang results, the baryon density, and the leptonic asymmetry. Phys. Rev. D **62**, 081301 (2000). https://doi.org/10.1103/PhysRevD.62.081301

19. J. Lesgourgues, A.R. Liddle, The lepton asymmetry: the last chance for a critical-density cosmology? Mon. Not. Roy. Astron. Soc. **327**, 1307 (2001). https://doi.org/10.1046/j.1365-8711.2001.04833.x

20. V. Simha, G. Steigman, Constraining the universal lepton asymmetry. JCAP **0808**, 011 (2008). https://doi.org/10.1088/1475-7516/2008/08/011

21. G. Mangano, G. Miele, S. Pastor, O. Pisanti, S. Sarikas, Constraining the cosmic radiation density due to lepton number with big bang nucleosynthesis. JCAP **1103**, 035 (2011). https://doi.org/10.1088/1475-7516/2011/03/035

22. G. Mangano, G. Miele, S. Pastor, O. Pisanti, S. Sarikas, Updated BBN bounds on the cosmological lepton asymmetry for non-zero θ_{13}. Phys. Lett. B **708**, 1–5 (2012). https://doi.org/10.1016/j.physletb.2012.01.015

23. E. Castorina, U. Franca, M. Lattanzi, J. Lesgourgues, G. Mangano, A. Melchiorri, S. Pastor, Cosmological lepton asymmetry with a nonzero mixing angle θ_{13}. Phys. Rev. D **86**, 023517 (2012). https://doi.org/10.1103/PhysRevD.86.023517

24. I.M. Oldengott, D.J. Schwarz, Improved constraints on lepton asymmetry from the cosmic microwave background. EPL **119**(2), 29001 (2017). https://doi.org/10.1209/0295-5075/119/29001

25. A.D. Dolgov, S.H. Hansen, S. Pastor, S.T. Petcov, G.G. Raffelt, D.V. Semikoz, Cosmological bounds on neutrino degeneracy improved by flavor oscillations. Nucl. Phys. B **632**, 363–382 (2002). https://doi.org/10.1016/S0550-3213(02)00274-2

26. Y.Y.Y. Wong, Analytical treatment of neutrino asymmetry equilibration from flavor oscillations in the early universe. Phys. Rev. D **66**, 025015 (2002). https://doi.org/10.1103/PhysRevD.66.025015

27. K.N. Abazajian, J.F. Beacom, N.F. Bell, Stringent constraints on cosmological neutrino antineutrino asymmetries from synchronized flavor transformation. Phys. Rev. D **66**, 013008 (2002). https://doi.org/10.1103/PhysRevD.66.013008

28. G. Barenboim, W.H. Kinney, W.I. Park, Resurrection of large lepton number asymmetries from neutrino flavor oscillations. Phys. Rev. D **95**(4), 043506 (2017). https://doi.org/10.1103/PhysRevD.95.043506

29. J.A. Harvey, E.W. Kolb, Grand unified theories and the lepton number of the universe. Phys. Rev. D **24**, 2090 (1981). https://doi.org/10.1103/PhysRevD.24.2090

30. P. Langacker, G. Segre, S. Soni, Majorana neutrinos, nucleosynthesis, and the lepton asymmetry of the universe. Phys. Rev. D **26**, 3425 (1982). https://doi.org/10.1103/PhysRevD.26.3425

31. B.A. Campbell, S. Davidson, J.R. Ellis, K.A. Olive, On the baryon, lepton flavor and right-handed electron asymmetries of the universe. Phys. Lett. B **297**, 118–124 (1992). https://doi.org/10.1016/0370-2693(92)91079-O
32. J. Liu, G. Segre, Baryon asymmetry of the universe and large lepton asymmetries. Phys. Lett. B **338**, 259–262 (1994). https://doi.org/10.1016/0370-2693(94)91375-7
33. R. Foot, M.J. Thomson, R.R. Volkas, Large neutrino asymmetries from neutrino oscillations. Phys. Rev. D **53**, R5349–R5353 (1996). https://doi.org/10.1103/PhysRevD.53.R5349
34. R. Foot, R.R. Volkas, Studies of neutrino asymmetries generated by ordinary sterile neutrino oscillations in the early universe and implications for big bang nucleosynthesis bounds. Phys. Rev. D **55**, 5147–5176 (1997). https://doi.org/10.1103/PhysRevD.55.5147
35. A. Casas, W.Y. Cheng, G. Gelmini, Generation of large lepton asymmetries. Nucl. Phys. **B538**, 297–308 (1999). https://doi.org/10.1016/S0550-3213(98)00606-3
36. B. Bajc, A. Riotto, G. Senjanovic, Large lepton number of the universe and the fate of topological defects. Phys. Rev. Lett. **81**, 1355–1358 (1998). https://doi.org/10.1103/PhysRevLett.81.1355
37. A.D. Dolgov, S.H. Hansen, S. Pastor, D.V. Semikoz, Neutrino oscillations in the early universe: how large lepton asymmetry can be generated? Astropart. Phys. **14**, 79–90 (2000). https://doi.org/10.1016/S0927-6505(00)00111-0
38. A. Sorri, Physical origin of 'chaoticity' of neutrino asymmetry. Phys. Lett. B **477**, 201–207 (2000). https://doi.org/10.1016/S0370-2693(00)00203-3
39. R. Buras, D.V. Semikoz, Lepton asymmetry creation in the early universe. Astropart. Phys. **17**, 245–261 (2002). https://doi.org/10.1016/S0927-6505(01)00155-4
40. R. Buras, D.V. Semikoz, Maximum lepton asymmetry from active sterile neutrino oscillations in the early universe. Phys. Rev. D **64**, 017302 (2001). https://doi.org/10.1103/PhysRevD.64.017302
41. K. Kainulainen, A. Sorri, Oscillation induced neutrino asymmetry growth in the early universe. JHEP **02**, 020 (2002). https://doi.org/10.1088/1126-6708/2002/02/020
42. M. Yamaguchi, Generation of cosmological large lepton asymmetry from a rolling scalar field. Phys. Rev. D **68**, 063507 (2003). https://doi.org/10.1103/PhysRevD.68.063507
43. L. Bento, F.C. Santos, Neutrino helicity asymmetries in leptogenesis. Phys. Rev. D **71**, 096001 (2005). https://doi.org/10.1103/PhysRevD.71.096001
44. J. Aasi et al., Advanced LIGO. Class. Quant. Grav. **32**, 074001 (2015). https://doi.org/10.1088/0264-9381/32/7/074001
45. B.P. Abbott et al., Binary black hole mergers in the first advanced LIGO observing run. Phys. Rev. **X6**(4), 041015 (2016). https://doi.org/10.1103/PhysRevX.6.041015
46. B.P. Abbott et al., The rate of binary black hole mergers inferred from advanced LIGO observations surrounding GW150914. Astrophys. J. **833**(1), L1 (2016). https://doi.org/10.3847/2041-8205/833/1/L1
47. A. Lue, L.M. Wang, M. Kamionkowski, Cosmological signature of new parity violating interactions. Phys. Rev. Lett. **83**, 1506–1509 (1999). https://doi.org/10.1103/PhysRevLett.83.1506
48. M. Novello, H.J. Mosquera Cuesta, V.A. DeLorenci, Birefringence of gravitational waves, in *Recent Developments in Theoretical and Experimental General Relativity, Gravitation and Relativistic Field Theories. Proceedings, 9th Marcel Grossmann Meeting, MG'9, Rome, Italy, July 2–8, 2000. Pts. A–C* (2000), pp. 1092–1095
49. S. Alexander, J. Martin, Birefringent gravitational waves and the consistency check of inflation. Phys. Rev. D **71**, 063526 (2005). https://doi.org/10.1103/PhysRevD.71.063526
50. H.S.S. Alexander, Inflationary birefringence and baryogenesis. Int. J. Mod. Phys. **D25**(11), 1640013 (2016). https://doi.org/10.1142/S0218271816400137
51. R. Jackiw, S.Y. Pi, Chern-Simons modification of general relativity. Phys. Rev. D **68**, 104012 (2003). https://doi.org/10.1103/PhysRevD.68.104012
52. S. Alexander, N. Yunes, Chern-Simons modified general relativity. Phys. Rept. **480**, 1–55 (2009). https://doi.org/10.1016/j.physrep.2009.07.002
53. P.A. Seoane et al., The gravitational universe (2013), arXiv:1305.5720

54. C. Caprini et al., Science with the space-based interferometer eLISA. II: gravitational waves from cosmological phase transitions. JCAP **1604**(04), 001 (2016). https://doi.org/10.1088/1475-7516/2016/04/001

55. K. Ichiki, M. Yamaguchi, J.I. Yokoyama, Lepton asymmetry in the primordial gravitational wave spectrum. Phys. Rev. D **75**, 084017 (2007). https://doi.org/10.1103/PhysRevD.75.084017

56. P. Singh, B. Mukhopadhyay, Gravitationally induced neutrino asymmetry. Mod. Phys. Lett. A **18**, 779–785 (2003). https://doi.org/10.1142/S0217732303009691

57. A. Denner, H. Eck, O. Hahn, J. Kublbeck, Feynman rules for fermion number violating interactions. Nucl. Phys. **B387**, 467–481 (1992). https://doi.org/10.1016/0550-3213(92)90169-C

58. D. Burns, A. Pilaftsis, Matter quantum corrections to the graviton self-energy and the Newtonian potential. Phys. Rev. D **91**(6), 064047 (2015). https://doi.org/10.1103/PhysRevD.91.064047

59. D. Anselmi, A Note on the dimensional regularization of the standard model coupled with quantum gravity. Phys. Lett. B **596**, 90–95 (2004). https://doi.org/10.1016/j.physletb.2004.06.089

60. M. Gomes, T. Mariz, J.R. Nascimento, E. Passos, AYu. Petrov, A.J. da Silva, On the ambiguities in the effective action in Lorentz-violating gravity. Phys. Rev. D **78**, 025029 (2008). https://doi.org/10.1103/PhysRevD.78.025029

61. A.N. Redlich, L.C.R. Wijewardhana, Induced Chern-Simons terms at high temperatures and finite densities. Phys. Rev. Lett. **54**, 970 (1985). https://doi.org/10.1103/PhysRevLett.54.970

62. J.G. McCarthy, A. Wilkins, Induced Chern-Simons terms. Phys. Rev. D **58**, 085007 (1998). https://doi.org/10.1103/PhysRevD.58.085007

63. M. Joyce, M. Shaposhnikov, Primordial magnetic fields, right-handed electrons, and the Abelian anomaly. Phys. Rev. Lett. **79**, 1193–1196 (1997). https://doi.org/10.1103/PhysRevLett.79.1193

64. R. Jackiw, V.A. Kostelecky, Radiatively induced Lorentz and CPT violation in electrodynamics. Phys. Rev. Lett. **82**, 3572–3575 (1999). https://doi.org/10.1103/PhysRevLett.82.3572

65. T. Mariz, J.R. Nascimento, E. Passos, R.F. Ribeiro, Chern-Simons-like action induced radiatively in general relativity. Phys. Rev. D **70**, 024014 (2004). https://doi.org/10.1103/PhysRevD.70.024014

66. M. Dvornikov, V.B. Semikoz, Instability of magnetic fields in electroweak plasma driven by neutrino asymmetries. JCAP **1405**, 002 (2014). https://doi.org/10.1088/1475-7516/2014/05/002

67. Y. Akamatsu, N. Yamamoto, Chiral plasma instabilities. Phys. Rev. Lett. **111**, 052002 (2013). https://doi.org/10.1103/PhysRevLett.111.052002

68. J.C.C. Felipe, A.R. Vieira, A.L. Cherchiglia, A.P. Baêta, A.B. Scarpelli, M. Sampaio, Arbitrariness in the gravitational Chern-Simons-like term induced radiatively. Phys. Rev. D **89**(10), 105034 (2014). https://doi.org/10.1103/PhysRevD.89.105034

69. M.M. Anber, E. Sabancilar, Chiral gravitational waves from chiral fermions. Phys. Rev. D **96**(2), 023501 (2017). https://doi.org/10.1103/PhysRevD.96.023501

70. M.B. Green, J.H. Schwarz, Anomaly cancellation in supersymmetric D = 10 gauge theory and superstring theory. Phys. Lett. B **149**, 117–122 (1984). https://doi.org/10.1016/0370-2693(84)91565-X

71. T.L. Smith, A.L. Erickcek, R.R. Caldwell, M. Kamionkowski, The effects of Chern-Simons gravity on bodies orbiting the earth. Phys. Rev. D **77**, 024015 (2008). https://doi.org/10.1103/PhysRevD.77.024015

72. N. Yunes, D.N. Spergel, Double binary pulsar test of dynamical Chern-Simons modified gravity. Phys. Rev. D **80**, 042004 (2009). https://doi.org/10.1103/PhysRevD.80.042004

73. Y. Ali-Haimoud, Revisiting the double-binary-pulsar probe of non-dynamical Chern-Simons gravity. Phys. Rev. D **83**, 124050 (2011). https://doi.org/10.1103/PhysRevD.83.124050

74. S. Dyda, E.E. Flanagan, M. Kamionkowski, Vacuum instability in Chern-Simons gravity. Phys. Rev. D **86**, 124031 (2012). https://doi.org/10.1103/PhysRevD.86.124031

75. S. Alexander, L.S. Finn, N. Yunes, A gravitational-wave probe of effective quantum gravity. Phys. Rev. D **78**, 066005 (2008). https://doi.org/10.1103/PhysRevD.78.066005
76. S.M. Carroll, M. Hoffman, M. Trodden, Can the dark energy equation-of-state parameter w be less than -1? Phys. Rev. D **68**, 023509 (2003). https://doi.org/10.1103/PhysRevD.68.023509
77. J.M. Cline, S. Jeon, G.D. Moore, The phantom menaced: constraints on low-energy effective ghosts. Phys. Rev. D **70**, 043543 (2004). https://doi.org/10.1103/PhysRevD.70.043543

Chapter 6
Concluding Remarks and Outlook

The Standard Models of Particle Physics and Cosmology have been highly successful at describing and reproducing the observed dynamics and properties of the Universe, but they are incomplete. There are still many mysteries of nature yet to be solved, for which new physics beyond the standard paradigms is required. In this thesis we have tried to propose solutions for some of these problems by considering particle physics ideas, specifically in relation to quantum anomalies, to early universe cosmology. This was motivated by the knowledge of the strong intertwining of cosmological and particle dynamics at very early times, when the microscopic dynamics of the fundamental particles directly dictated the evolution of the universe. Through considering a combination of observables from terrestrial collider searches and cosmological observables it may be possible to piece together the answers to many of the open questions of our universe.

In Chap. 2, we proposed a new class of natural inflation models which provided a solution to the hierarchy problem through a hidden scale invariance realised through the introduction of a dilaton field [1]. Given the scale invariant symmetry of the theory, the inflationary potential naturally contains a flat direction in the classical limit, which is lifted by quantum corrections. Thus inflation can naturally, without fine-tuning, proceed when the inflaton field evolves along this direction. We find that in the conformal limit, the inflaton potential is linear, which gives predictions in agreement with observations. Therefore, this model provides a successful inflationary scenario within which a solution to the hierarchy problem of the Standard Model can be found.

Chapter 3 presented an unorthodox mechanism for the origin of the matter-antimatter asymmetry as well as dark matter; one that acts during the inflationary epoch [2, 3]. This mechanism for cogenesis involved the introduction of an anomalous gauge interaction and sterile fermion to the Standard Model. The anomalies associated with the new gauge field provided the X charge violation, and the corresponding counter terms violated C and CP in the cosmological setting, while the inflationary epoch provided the push out-of-equilibrium. It was found that this scenario

© Springer International Publishing AG, part of Springer Nature 2018
N. D. Barrie, *Cosmological Implications of Quantum Anomalies*,
Springer Theses, https://doi.org/10.1007/978-3-319-94715-0_6

for cogenesis can successfully reproduce the observed values of the baryon asymmetry and dark matter abundance for the two possible cases considered—gauged B and gauged $B - L$ charge—for certain parameter spaces. The general mechanism for cogenesis developed here could be applied to more complex models involving other or extra anomalous gauge symmetries and additional sterile or non-sterile fermionic states. It is possible that these additions could lead to a lessening of the parameter constraints imposed by the observed matter-antimatter asymmetry, through extra contributions to the luminous matter generation. Given that this model involves the introduction of new gauge bosons and a dark sector to the Standard Model, this mechanism could have potential avenues for experimental investigation at terrestrial collider searches and direct detection experiments. Therefore, it may be possible to utilise both terrestrial and cosmological measurements to constrain this mechanism.

In Chap. 4, we presented a model for Baryogenesis during reheating that utilises the Ratchet Mechanism [4]. We introduced a theory containing two fundamental scalars, an inflaton consistent with the Starobinsky inflationary mechanism, and a complex scalar baryon with a symmetric potential; with the two scalars interacting via a derivative coupling. The scalar baryon potential violates B, and the violation of C and CP is introduced by the derivative coupling interaction. The push out-of-equilibrium in this mechanism is provided by the reheating epoch, which is caused by the coherent oscillation of the inflaton in its potential. In order for a non-zero baryon number density to be produced, driven motion must be induced in the phase of the complex scalar baryon. The inflaton-scalar baryon system was found to act analogously to a forced pendulum, with driven motion achieved near the end of reheating for parameters consistent with the Sweet Spot Condition. This result implied a high reheating temperature as a generic requirement of our model. Further analysis of this mechanism could provide interesting cosmological phenomenology beyond the reheating epoch through the decays and interactions of the baryonic scalar.

Chapter 5 discussed a novel way to utilise gravitational waves to illuminate the properties of the illusive Cosmic Neutrino Background [5]. We explicitly calculated the parity odd part of the graviton polarization tensor in the presence of a lepton asymmetric Cosmic Neutrino Background, which generates a gravitational Chern-Simons term in the effective action, in the vanishing neutrino mass limit. The induced Chern-Simons term causes birefringent behaviour in gravitational wave propagation leading to an enhancement or suppression of the gravitational wave amplitude depending on the polarisation. While this effect is negligibly small for local sources, we demonstrated that it could be sizeable for gravitational waves produced in the very early universe with a momenta $|\mathbf{k}| \sim H$. We also argued that a relic neutrino background with non-zero lepton number exhibits gravitational instabilities that are related to the gravity-lepton number mixed anomaly. The induced negative energy modes, lead to a rapid decay of a vacuum state into photons and gravitons, from which we could derive observational bounds. From the constraints on the radiation energy density today, we were able to obtain an interesting bound on the neutrino degeneracy parameter. We believe that the findings reported in this Chapter will prove to be useful for further understanding of the properties of the Cosmic Neutrino Background and allow constraints to be placed on particle physics models that generate a lepton asymmetry.

Being able to constrain or obtain indirect measurements of the size of the lepton asymmetry would help illuminate the properties of the neutrinos and possibly the mechanism for Baryogenesis. This work explored the exciting new possibility of using gravitational wave phenomena to uncover information about the properties of the fundamental particles of nature.

In this thesis, we sought to demonstrate the importance of particle physics in the evolution of the early universe, and some of the interesting ways in which this could be explored; predominantly through the consideration of quantum anomalies. Increased investigation into such phenomena, as well as other particle physics applications to cosmology, will undoubtedly further increase our understanding of nature. It is clear that this approach, in concert with terrestrial particle physics phenomenology and experimental searches, is the clear way forward in our endeavour to understand the fundamental nature of the world around us.

References

1. N.D. Barrie, A. Kobakhidze, S. Liang, Natural inflation with hidden scale invariance. Phys. Lett. B **756**, 390–393 (2016a). https://doi.org/10.1016/j.physletb.2016.03.056
2. N.D. Barrie, A. Kobakhidze, Inflationary baryogenesis in a model with gauged baryon number. JHEP **09**, 163 (2014). https://doi.org/10.1007/JHEP09(2014)163
3. N.D. Barrie, A. Kobakhidze, Generating luminous and dark matter during inflation. Mod. Phys. Lett. A **32**(14), 1750087 (2017a). https://doi.org/10.1142/S0217732317500870
4. K. Bamba, N.D. Barrie, A. Sugamoto, T. Takeuchi, K. Yamashita, Ratchet baryogenesis with an analogy to the forced pendulum (2016), arXiv:1610.03268
5. N.D. Barrie, A. Kobakhidze, Gravitational instabilities of the cosmic neutrino background with non-zero lepton number. Phys. Lett. B **772**, 459–463 (2017b). https://doi.org/10.1016/j.physletb.2017.07.012

Appendix A
Baryon and Lepton Number Anomalies in the Standard Model

A.1 Baryon Number Anomalies

The introduction of a gauged baryon number leads to the inclusion of quantum anomalies in the theory, refer to Fig. 1.2. The anomalies, for the baryonic current, are given by the following,

For $SU(3)^2 U(1)_B$,

$$\mathcal{A}_1(SU(3)^2 U(1)_B) = Tr[\lambda^a \lambda^b B] = 3 \times \frac{3}{2} \left(\sum_{left} B_i - \sum_{right} B_i \right) = 0. \quad \text{(A.1)}$$

For $SU(2)^2 U(1)_B$,

$$\mathcal{A}_2(SU(2)^2 U(1)_B) = Tr[\tau^a \tau^b B] = \frac{3 \times 3}{2} B_Q = \frac{3}{2}. \quad \text{(A.2)}$$

For $U(1)_Y^2 U(1)_B$,

$$\mathcal{A}_3(U(1)_Y^2 U(1)_B) = Tr[YYB] = 3 \times 3(2Y_Q^2 B_Q - Y_u^2 B_u - Y_d^2 B_d) = -\frac{3}{2}. \quad \text{(A.3)}$$

For $U(1)_B^2 U(1)_Y$,

$$\mathcal{A}_4(U(1)_B^2 U(1)_Y) = Tr[BBY] = 3 \times 3(2B_Q^2 Y_Q - B_u^2 Y_u - B_d^2 Y_d) = 0. \quad \text{(A.4)}$$

For $U(1)_B^3$,

$$\mathcal{A}_5(U(1)_B^3) = Tr[BBB] = 3 \times 3(2B_Q^3 - B_u^3 - B_d^3) = 0. \quad \text{(A.5)}$$

© Springer International Publishing AG, part of Springer Nature 2018
N. D. Barrie, *Cosmological Implications of Quantum Anomalies*,
Springer Theses, https://doi.org/10.1007/978-3-319-94715-0

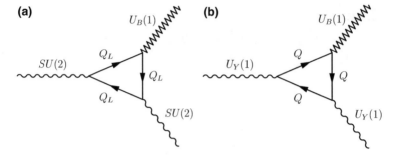

Fig. A.1 1-Loop corrections to **a** $SU(2)^2 U(1)_B$, where the loop contains only left-handed quarks, and **b** $U(1)_Y^2 U(1)_B$ where the loop contains only quarks

For $U(1)_B$,

$$\mathcal{A}_6(U(1)_B) = Tr[B] = 3 \times 3(2B_Q - B_u - B_d) = 0, \tag{A.6}$$

where the factor of 3×3 is a result of there being three generations of quarks and three colours for each quark. The δ^{ab} terms are not included in the anomalies (Fig. A.1).

A.2 Lepton Number Anomalies

When introducing right handed neutrinos into the SM the quantum anomalies for a gauged lepton number, or leptonic current, are the following,

For $SU(3)^2 U(1)_L$,

$$\mathcal{A}_1(SU(3)^2 U(1)_L) = Tr[\lambda^a \lambda^b L] = \frac{3}{2}\left(\sum_{left} L_i - \sum_{right} L_i\right) = 0. \tag{A.7}$$

For $SU(2)^2 U(1)_L$,

$$\mathcal{A}_2(SU(2)^2 U(1)_L) = Tr[\tau^a \tau^b L] = \frac{3}{2}L_L = \frac{3}{2}. \tag{A.8}$$

For $U(1)_Y^2 U(1)_L$,

$$\mathcal{A}_3(U(1)_Y^2 U(1)_L) = Tr[YYL] = 3(2Y_L^2 L_L - Y_e^2 L_e - Y_\nu^2 L_\nu) = -\frac{3}{2}. \tag{A.9}$$

For $U(1)_L^2 U(1)_Y$,

$$\mathcal{A}_4(U(1)_L^2 U(1)_Y) = Tr[LLY] = 3(2L_L^2 Y_L - L_e^2 Y_e - L_\nu^2 Y_\nu) = 0. \qquad \text{(A.10)}$$

For $U(1)_L^3$,

$$\mathcal{A}_5(U(1)_L^3) = Tr[LLL] = 3(2L_L^3 - L_e^3 - L_\nu^3) = 0. \qquad \text{(A.11)}$$

For $U(1)_L$,

$$\mathcal{A}_6(U(1)_L) = Tr[L] = 3(2L_L - L_e - L_\nu) = 0. \qquad \text{(A.12)}$$

If the right handed neutrinos are not included in the SM, $\mathcal{A}_5$ and $\mathcal{A}_6$ will be non-zero. That is, $\mathcal{A}_5 = 3$ and $\mathcal{A}_6 = 3$, where $\mathcal{A}_6$ is to the graviton-lepton anomaly.

A.3   Mixed Gauged Baryon and Lepton Number Anomalies

If these two gauge groups are introduced then the interactions between the leptonic and baryonic currents must also be anomaly free,
For $U(1)_B^2 U(1)_L$,

$$\mathcal{A}(U(1)_B^2 U(1)_L) = Tr[BBL] = 0. \qquad \text{(A.13)}$$

For $U(1)_B U(1)_L^2$,

$$\mathcal{A}(U(1)_L^2 U(1)_B) = Tr[LLB] = 0. \qquad \text{(A.14)}$$

For $U(1)_B U(1)_L U(1)_Y$,

$$\mathcal{A}(U(1)_B U(1)_L U(1)_L) = Tr[BLY] = 0. \qquad \text{(A.15)}$$

These will only be non-zero if fermions such as leptoquarks are added to the SM. There are no fermions in the SM which can couple to both a leptophobic gauge boson and a leptophillic gauge boson.

Some recent models have introduced leptoquarks along with gauged baryon and lepton number symmetries into the SM [1, 2]. To ensure that these mixed interactions don't lead to new gauge anomalies, the number of types of leptoquarks and the quantum numbers they carry are such that these quantum corrections remain zero. They can also be used to cancel the gauge anomalies that are also present with these gauge bosons in combination with the SM gauge fields.

References

1. P.V. Dong, H.N. Long, A simple model of gauged lepton and baryon charges. Phys. Int. **6**(1), 23–32 (2010). https://doi.org/10.3844/pisp.2015.23.32
2. M. Duerr, P.F. Perez, M.B. Wise, Gauge theory for baryon and lepton numbers with leptoquarks. Phys. Rev. Lett. **110**, 231801 (2013). https://doi.org/10.1103/PhysRevLett.110.231801

Appendix B
Further Details of Chap. 3 Calculations

B.1 F_+ Coefficients, Eq. (3.20)

Matching superhorizon modes with the plane waves, we obtain the following relation,

$$C_1 = \frac{\Gamma\left(\frac{3-\Omega_k}{4}\right)}{2^{\frac{-1}{4}(1-\Omega_k)}\sqrt{\pi}}\left(\frac{1}{\sqrt{2k}} - C_2 \frac{2^{\frac{-1}{4}(1+\Omega_k)}\sqrt{\pi}}{\Gamma\left(\frac{3+\Omega_k}{4}\right)}\right). \tag{B.1}$$

The Wronskian normalisation implies:

$$\sqrt{\frac{2}{\Omega_k}}C_1 C_2 \sin\left(\frac{\pi}{4}(1+\Omega_k)\right) + C_2^2 \sqrt{\frac{\pi}{\Omega_k}}\frac{1}{\Gamma\left(\frac{1+\Omega_k}{2}\right)} = \frac{1}{2k}. \tag{B.2}$$

Solving the above conditions we find that the coefficients for the F_+ modes are,

$$C_1 = \frac{2^{-\frac{1}{4}(1+\Omega_k)}\Gamma\left(\frac{3-\Omega_k}{4}\right)}{\sqrt{\pi k}} - \frac{2^{-\frac{1}{2}(\Omega_k+3)}\Gamma\left(\frac{1+\Omega_k}{4}\right)\Gamma\left(\frac{3-\Omega_k}{4}\right)}{\Gamma\left(\frac{3+\Omega_k}{4}\right)}\sqrt{\frac{\Omega_k}{\pi k}}, \tag{B.3}$$

and

$$C_2 = \frac{\Gamma\left(\frac{1+\Omega_k}{4}\right)}{2\sqrt{2\pi}}\sqrt{\frac{\Omega_k}{k}} = \frac{\Gamma\left(\frac{1+\Omega_k}{4}\right)}{2\sqrt{2\pi}}\left(\frac{k}{\kappa}\right)^{\frac{1}{4}}. \tag{B.4}$$

B.2 F_- Coefficients, Eq. (3.21)

Similarly as above, we obtain the following relations from the matching,

$$C_4 = \frac{\Gamma\left(\frac{3-i\Omega_k}{4}\right)}{2^{\frac{-1}{4}(1-i\Omega_k)}\sqrt{\pi}}\left(\frac{1}{\sqrt{2k}} - C_3 \frac{2^{\frac{-1}{4}(1+i\Omega_k)}\sqrt{\pi}}{\Gamma\left(\frac{3+i\Omega_k}{4}\right)}\right), \tag{B.5}$$

© Springer International Publishing AG, part of Springer Nature 2018
N. D. Barrie, *Cosmological Implications of Quantum Anomalies*,
Springer Theses, https://doi.org/10.1007/978-3-319-94715-0

and the Wronskian normalisation,

$$C_3^2 + |C_4|^2 + 2C_3 e^{\frac{-\pi\Omega_k}{4}} \sqrt{2\pi} \operatorname{Im}\left(\frac{\sqrt{i}\, C_4^*}{\Gamma\left(\frac{1+i\Omega_k}{2}\right)}\right) = \frac{e^{\frac{-\pi\Omega_k}{4}}}{k}\sqrt{\frac{\Omega_k}{2}}. \tag{B.6}$$

These two equations determine the coefficients for the F_- modes,

$$C_3 = \frac{1}{2\sqrt{2k}\,P(k)}\left(\sqrt{\Omega_k}\,e^{-\frac{\pi\Omega_k}{4}} - \frac{1}{\pi}\left|\Gamma\left(\frac{3-i\Omega_k}{4}\right)\right|^2\right), \tag{B.7}$$

and

$$C_4 = \frac{\Gamma\left(\frac{3-i\Omega_k}{4}\right)}{2^{\frac{-1}{4}(1-i\Omega_k)}\sqrt{2\pi k}}\left(1 - \frac{\sqrt{\pi}}{2^{\frac{1}{4}(5+i\Omega_k)}\,P(k)\,\Gamma\left(\frac{3+i\Omega_k}{4}\right)}\left(\sqrt{\Omega_k}\,e^{-\frac{\pi\Omega_k}{4}} - \frac{1}{\pi}\left|\Gamma\left(\frac{3-i\Omega_k}{4}\right)\right|^2\right)\right), \tag{B.8}$$

where

$$P(k) = \frac{2^{3/4}}{\sqrt{\pi}}\left(2\pi e^{-\frac{\pi\Omega_k}{4}}\operatorname{Im}\left[\frac{\sqrt{i}}{2^{\frac{i\Omega_k}{4}}\Gamma\left(\frac{1+i\Omega_k}{4}\right)}\right] - \operatorname{Re}\left[\frac{\Gamma\left(\frac{3-i\Omega_k}{4}\right)}{2^{\frac{i\Omega_k}{4}}}\right]\right). \tag{B.9}$$

Appendix C
Further Details of Chap. 5 Calculations

C.1 Dimensional Regularisation Integrals and Useful Relations

The following dimensional regularisation integrals were utilised in Chap. 5,

$$i \int \frac{d^N p}{(2\pi)^N} \frac{1}{(p^2 - m^2)^2} = -\frac{1}{16\pi^2} \left[\frac{4\pi^2 \lambda^2}{M^2} \right]^\epsilon \Gamma(\epsilon), \tag{C.1}$$

$$i \int \frac{d^N p}{(2\pi)^N} \frac{1}{(p^2 - m^2)^3} = \frac{1}{32\pi^2} \left[\frac{4\pi^2 \lambda^2}{M^2} \right]^\epsilon \frac{\Gamma(1 + \epsilon)}{M^2}, \tag{C.2}$$

$$i \int \frac{d^N p}{(2\pi)^N} \frac{p_\mu p_\nu}{(p^2 - m^2)^2} = \frac{1}{32\pi^2} \left[\frac{4\pi^2 \lambda^2}{M^2} \right]^\epsilon M^2 \Gamma(\epsilon - 1) g_{\mu\nu}, \tag{C.3}$$

$$i \int \frac{d^N p}{(2\pi)^N} \frac{p_\mu p_\nu}{(p^2 - m^2)^3} = -\frac{1}{64\pi^2} \left[\frac{4\pi^2 \lambda^2}{M^2} \right]^\epsilon \Gamma(\epsilon) g_{\mu\nu}, \tag{C.4}$$

$$i \int \frac{d^N p}{(2\pi)^N} \frac{p_\mu p_\nu p_\rho p_\sigma}{(p^2 - m^2)^3} = \frac{1}{128\pi^2} \left[\frac{4\pi^2 \lambda^2}{M^2} \right]^\epsilon M^2 \Gamma(\epsilon - 1)(g_{\mu\nu} g_{\rho\sigma} + g_{\mu\sigma} g_{\nu\rho} + g_{\mu\rho} g_{\nu\sigma}). \tag{C.5}$$

Some other useful relations are,

$$Tr(\gamma_\mu \gamma_\alpha \gamma_\rho \gamma_\beta \gamma^5) = -4i \varepsilon_{\mu\alpha\rho\beta}, \tag{C.6}$$

© Springer International Publishing AG, part of Springer Nature 2018
N. D. Barrie, *Cosmological Implications of Quantum Anomalies*,
Springer Theses, https://doi.org/10.1007/978-3-319-94715-0

$$Tr(\gamma_\mu(\not p + \not k + m)\gamma_\rho\sigma_{\alpha\beta}\gamma^5(\not p + m)) =$$
$$4\{\varepsilon_{\mu\rho\alpha\beta}[m^2 - p^2 - (kp)] - k^\lambda[\varepsilon_{\alpha\beta\rho\lambda}P_\mu - \varepsilon_{\alpha\beta\mu\lambda}P_\rho]\}, \qquad \text{(C.7)}$$

$$Tr(\gamma_\rho(\not p - \not k + m)\gamma_\mu\sigma_{\alpha\beta}\gamma^5(\not p + m)) =$$
$$4\{\varepsilon_{\mu\rho\alpha\beta}[p^2 - m^2 - (kp)] - k^\lambda[\varepsilon_{\alpha\beta\rho\lambda}P_\mu - \varepsilon_{\alpha\beta\mu\lambda}P_\rho]\}. \qquad \text{(C.8)}$$

Upon taking $\epsilon \to 0$ the following are obtained,

$$\Gamma(1+\epsilon)|_{\epsilon\to0} \simeq 1, \quad \Gamma(\epsilon)|_{\epsilon\to0} \simeq \frac{1}{\epsilon} - \gamma, \quad \Gamma(\epsilon-1)|_{\epsilon\to0} \simeq -\frac{1}{\epsilon} + \gamma - 1, \quad \text{(C.9)}$$

$$\eta_{\mu\nu}\eta^{\mu\nu} \simeq 4 - 2\epsilon, \qquad \text{(C.10)}$$

$$\left[\frac{4\pi\lambda^2}{M^2}\right]^\epsilon|_{\epsilon\to0} \simeq 1 + \epsilon\ln\left(\frac{4\pi\lambda^2}{M^2}\right). \qquad \text{(C.11)}$$